SCIENCE AND SOCIETY

THE MEANING AND IMPORTANCE OF
SCIENTIFIC METHOD

SCIENCE AND SOCIETY

THE MEANING AND IMPORTANCE OF
SCIENTIFIC METHOD

Michael Bassey

 UNIVERSITY OF LONDON PRESS LTD

SBN 340 06975 9

Copyright © 1968 Michael Bassey
Illustrations copyright © 1968 University of London Press Ltd
All rights reserved. No part of this publication may be reproduced
or transmitted in any form or by any means, electronic or mechanical
including photocopy, recording, or any information storage and
retrieval system, without permission in writing from the publisher.

University of London Press Ltd
St Paul's House, Warwick Lane, London EC4

Printed in Great Britain by
Neill and Co. Ltd, Edinburgh

Contents

Preface 7
Introduction 9

Chapter 1 PROBLEM SOLVING 11

Chapter 2 FACTS AND VALUES 26

Chapter 3 SCIENTIFIC METHOD 48

Chapter 4 SOME CONTEMPORARY SOCIAL PROBLEMS
 (i) Cigarette smoking and lung cancer 69
 (ii) Dental caries and the fluoridation of water supplies 77
 (iii) The Birmingham Dipped Headlights Experiment 82
 (iv) Learning to read by the Initial Teaching Alphabet 85
 (v) Crime and punishment 88

ENVOI 93
Further reading 94
References 95

Preface

The ideas in this book have slowly come together over a period of fifteen years and it is no longer possible to indicate the origin of every example cited, but I have tried to give the major sources in the table of references. One who is not listed there, but who deserves mention, is Professor Sir Ronald Nyholm; he first made me aware of the importance of scientific method in his undergraduate lectures at University College London. Apart from my indebtedness to the authors given in the table of references, and to those who have kindly given permission to reproduce illustrations, I would like to express my gratitude to Mr. S. Hoffman, Mr. G. P. Mabon and my wife, who have read through most of the manuscript and have made many valuable suggestions for its improvement. However I am responsible for any errors or obscurities that remain and I shall welcome suggestions which may remedy these from readers.

M. B.

For Mark and Katherine

May this make some contribution towards the solving of the immense problems which will face their generation.

Introduction

The future of man depends upon his skill in solving problems. Recent successes in problem solving have brought about enormous changes in agriculture, industry, transport, medicine and communications, which have considerably changed the pattern of human life and death. In particular the population of the world is now exploding, hurling problems like shrapnel in all directions.
As cynics have observed, if man does not control the population explosion, nature will. The other 'explosion' problem, 'the nuclear detergent which could clean life off the face of the earth', is equally needful of a rapid solution.
There are innumerable problems left in the wake of progress, which, if minor from the viewpoint of mankind's survival, are nevertheless of immense local consequence. National problems in Britain at the time of writing include: lung cancer, cigarette smoking and air pollution; children's teeth decay, the fluoridation of water and sweet-eating; road accidents, alcohol, speed limits and safety belts; juvenile crime, teenage riots, tranquilliser pills and violence on television; comprehensive schools and the educational opportunities of working-class and middle-class children.
It becomes increasingly important that the people who make

decisions, and in a democracy this is a large number, should be competent at problem solving. In order to do this they must be able to distinguish between facts and values, they must be able to search scientifically for Truth, and they must seek for, and use, a widely acceptable set of values based on love and beauty.

This book is a primer in problem solving. Although skill must depend largely on experience, the writer believes that an understanding of the approach which has come to be called 'scientific method', and an appreciation of its limitations, can be of considerable help in solving problems far removed from laboratory work. The trial and error process of thinking of likely hypotheses and of testing them to find which fit the facts, is a procedure used almost universally for simple everyday problems. If the key will not open the door we suggest that the wrong key is being used, or that the lock needs oiling, or that the door is bolted; and we soon work out which is the case. The method seems so obvious that many people dismiss discussion of it as trivial, failing to realise that it only seems trivial when the problem is trivial. Hypothesis testing is often neglected in the social, political and economic problems of society. In these complex issues people may confuse facts with values, or having isolated the facts and proposed a hypothesis, may cling to it without being prepared to test its validity. In arguing about capital punishment for murderers, some people are inclined to muddle the moral value issue of the sanctity of human life with the factual issue of the deterrent. Or having isolated the latter, they may stick to their opinion that capital punishment is a deterrent to murder, without being prepared to consider the factual evidence that is available. 'Don't confuse me with the facts; I have made up my mind,' is unfortunately a common remark.

In the popular image, scientists are white-coated men who shake test-tubes and peer into microscopes. Many have spent painstaking lives engaged in such pursuits, but the results of their work have been not only the immediate findings, but also the slow realisation that the scrupulous rules of intellectual honesty that the scientists have obeyed have a far wider application than chemistry, physics and biology. Scientific method cannot solve all the problems of the world, but it can make a useful contribution to solving many of them.

In the first chapter some of the different ways in which problems can be tackled are discussed. The terms 'fact' and 'value' are examined in some detail in the second chapter, and lead to discussion of the scientific method of tackling problems in the third chapter. Finally, in the fourth chapter, ideas developed earlier are applied to some contemporary social problems.

Chapter 1

Problem Solving

'The world is teeming with problems. Wherever man looks, he encounters some new problem—in his home life and in his job, in economics and in technology, in the arts and in the sciences. And some problems are very stubborn—they refuse to leave us in peace. They torture our thoughts, sometimes haunting us throughout the day and even robbing us of sleep at night.' (Max Planck)

Methods of Solving Problems

Science began, they say, when a man asked the first question. They are wrong; science began when a man first queried the answer to a question. Science is more than the asking of questions; it includes a preparedness to test the answers to questions, in an attempt to discover truth. Science should not be thought of as just chemistry, physics and the 'ologies; science should not be described as just a system of knowledge, for it is more than this: it is both the knowledge and the method by which the knowledge is being won. Here we are concerned with the description of science as a method of solving problems, as an intellectual tool, as a probe for exploring the unknown; and we are further concerned with the view that, of the two great realms of human experience—the realm of facts and the realm of values, science is only competent in the former. Science is a method of solving problems of fact.

To a young child the universe consists of his home and the

neighbouring streets. As he grows, his universe grows too. In an analogous way science first means to the child simple experiments with air and water, develops into chemistry and physics and biology, and only later becomes identified as an activity within the whole realm of facts. We wonder why iron rusts and we find that it rusts when it is wet. We become scientific in our approach when we ask whether water alone causes rusting. By testing the first answer to our question we come nearer to the truth. Science is a method of searching for truth.

The solving of problems is a perpetual human activity, extending from 'Where do we find the next meal?', to 'How was the universe made?'

A variety of methods of solving problems are used by mankind and it is helpful to consider these before examining scientific method in detail.

In *The Proper Study of Mankind*, Stuart Chase (1957) discusses six different methods of solving problems, and, although the list is not exclusive, it forms a useful basis for discussion.

Appeal to the Supernatural   *Appeal to Worldly Authority*
Use of Intuition   *Use of Logic*
Use of Common Sense   *Use of Scientific Method*

Appeal to the Supernatural

A problem that must have faced many elders of tribal societies in the past was succinctly expressed by Flanders and Swann (1957) in their song' The Reluctant Cannibal'. A young man and his father clash over the consumption of human flesh. 'Eating people is wrong', chants the son, and his father, struggling with the problem of how to convince his erring son, replies with an age-old solution: 'If the Juju hadn't meant us to eat people, he wouldn't have made them of meat.' The father's methodology can be described as *appeal to the supernatural*.

Appeal to supernatural agencies is a very primitive way of obtaining help or of explaining phenomena. As young children we hear of elves and fairies, of goblins and gremlins, of witches and wizards. Our milk teeth are encouraged to drop out at the appropriate age by the promise of a sixpence from the fairies. A message posted up the chimney will help Father Christmas to know what to bring when he climbs down. As we grow older we tend to lose our belief's in these supernatural beings and move into the realm of adult superstitions, which are less concerned with solving problems than with avoiding them. Don't walk under ladders, cross knives at the table, open umbrellas indoors, or travel on Friday the 13th. Wish when you eat your first of the season piece of Christmas pudding; if you see a black cat; if you get the large piece of the chicken's 'Y' bone on breaking it with a friend. These dealings with the supernatural seem natural enough to some people, and even those who laugh are inclined to throw salt over the left shoulder if they happen to spill any on the table.

The popular press gives regular space to astrological advice on how best to tackle personal problems, and presumably this must be associated with some supernatural force, since there is no obvious natural connection between the space co-ordinates of various planets and the time co-ordinates of birthdays. Some people take this very seriously. In 1962 a magistrate who was Deputy Lieutenant of his county and chairman of the county police force

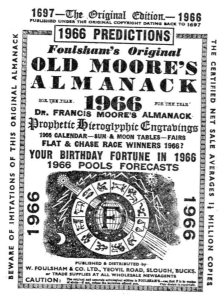

Figure 1 Problem solving by appeal to the supernatural. Old Moore's Almanack, published annually, gives predictions of future events based on the positions of the planets in the constellations of the Zodiac. Notice that the certified net sales are over one million. *Reproduced by kind permission of W. Foulsham and Co. Ltd., who own the copyright.*

told a Sunday newspaper, 'I look forward to the time when it will be standard practice to have available for magistrates an interpreted horoscope of every child charged with a serious offence.' (R1)

In the final analysis it would probably be shown that most people enjoy their superstitions and astrology rather than rely on them; these aspects of the supernatural provide more amusement than assistance. God alone remains as a supernatural power whose aid can be sought. Prayer is probably the only form of serious-minded problem-solving by appeal to the supernatural.

Appeal to Worldly Authority

In *Pseudodoxia Epidemica or Enquiries into Vulgar and Common Errors* (1646) Sir Thomas Browne drew attention to the belief of country folk that the badger has shorter legs on one side of its body, which peculiarity enables it to run easily along the slope of hills. Ask people today how they would check whether this is true, and a usual answer is that they would measure a few badgers, but on reflection they decide that it would be easier either to consult a naturalist, or to seek a book on badgers. They would appeal to worldly authority.

Solving problems by appeal to authority in the form of books or people is the simplest method of tackling most problems, provided that someone has dealt competently with the problem before, provided that we can find his solution and provided that we have confidence in his ability.

It is often said that it is as useful to know where to find a piece of information as to know the information itself. Libraries can contain vastly more knowledge than one man's mind and the written word is less likely to be in error than the memory. But it is important to be prepared for errors even in 'authoritative' works. For example, it has been known for some years that the number of human chromosomes is 46, but textbooks are still being published in which the erroneous number 48 is given. Few textbook writers have the time and opportunity to check all the original papers on which their writing is based, let alone repeat experimental work, and so errors tend to be handed on from one writer to the next. Figure 2 shows two illustrations of the strawberry plant from early herbals. One was drawn in the 7th century and the second about five hundred years later. It appears to be a copy of a copy of a copy. The 'runners' seem to change to thorns, and the plant is confused with the blackberry. (This is taken from G. Rattray Taylor, *The Science of Life*, 1963.)

Modern elementary textbook writers rarely have the opportunity

Figure 2 An example of errors handed from one writer to the next. The lower illustration of a strawberry plant is clearly a copy (probably several times removed) of the upper one, and repeats the error of grouping the leaves in fours and fives. There should, of course, be only three leaves to each stalk. The 'runners' are too numerous in the first drawing, and become confused with thorns in the second.

From a 7th century French herbal.

Reproduced by kind permission of Bibliotheek der Rijksuniversiteit te Leiden.

From a 12th century Rhenish herbal.

Reproduced by kind permission of Murhardsche und Landesbibliothek, Kassel.
(2 Ms. phys. 10.)

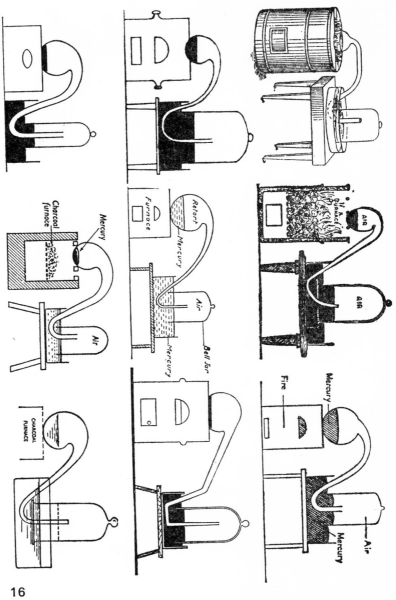

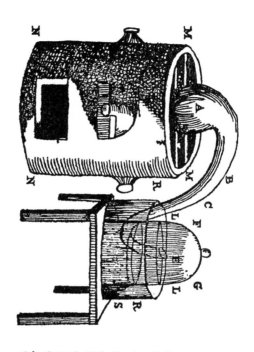

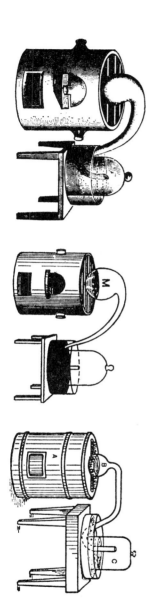

Figure 3 Diagrams of Lavoisier's dephlogisticated air experiment. The original diagram was prepared by Madame Lavoisier in 1789 for her husband's *Traité de Chimie*. Below it are examples of twentieth century copies from a number of elementary chemistry text-books. This illustrates the way in which authors rely on their predecessors for data, but introduce their own small alterations in order to express their own individuality. *After an illustration in* Technical Education, *December 1960.*

of reading original papers and so tend to reorganise the work of earlier writers. Figure 3 shows how changes in textbooks develop. Lavoisier's dephlogisticated air experiment is familiar to everyone who has studied elementary chemistry and it is usually illustrated in introductory texts. No chemical errors are involved in the alterations made by successive authors shown here, but it is curious to trace the changes that have been made from the original drawing, which was sketched by Madame Lavoisier.

In former days it was usual to believe that the older the authority the more reliable it would be. The writings of men like Aristotle (384–322 B.C.) and Galen (A.D. 130–200) were considered immutable for more than fifteen hundred years. It is said that in the 1620's a monk who noted sunspots was told by his superior, 'You are mistaken, my son. I have studied Aristotle and he nowhere mentions spots. Try changing your spectacles.' To challenge scientific authority required great courage. Galen wrote that the arteries and veins were unconnected and that the blood slowly ebbed and flowed in the two systems like the tide. In 1615 Harvey realised from experiments and logical argument that the blood circulates round the body, but he did not dare publish this discovery until 1628, when he was physician to Charles I. The writings of Francis Bacon (1561–1626) helped to dispel the reverence for authority. In 1605 he told the following fable:

'In the year of our Lord 1432, there arose a grievous quarrel among the brethren over the number of teeth in the mouth of a horse. For 13 days the disputation raged without ceasing. All the ancient books and chronicles were fetched out, and wonderful and ponderous erudition, such as was never before heard of in this region, was made manifest.
At the beginning of the 14th day, a youthful friar of goodly bearing asked his learned superiors for permission to add a word, and straightway, to the wonderment of the disputants, whose deep wisdom he sore vexed, he beseeched them to unbend in a manner coarse and unheard-of, and to look in the open mouth of a horse and find answer to their questionings. At this, their dignity being grievously hurt, they waxed exceedingly wroth; and, joining in a mighty uproar, they flew upon him and smote him hip and thigh, and cast him out forthwith. For, said they, surely Satan hath tempted this bold neophyte to declare unholy and unheard-of ways of finding truth contrary to all the teachings of the fathers. After many days more of grievous strife the dove of peace sat on the assembly, and they as one man, declaring the

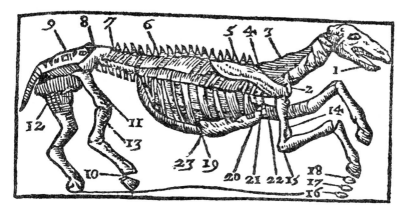

Figure 4 How many teeth has a horse? This drawing by G. B. Ferrari, a sixteenth century anatomist, would hardly help.

problem to be an everlasting mystery because of a grievous dearth of historical and theological evidence thereof, so ordered the same to be writ down." (R2)

Logic

Logic is a disciplined system of thinking by which conclusions drawn from factual statements, called premises, can be shown to be true or false. If steel is made in Sheffield, and Sheffield is in England, then steel is made in England. However, if steel is made in Sheffield, and knives are made of steel, it does not follow that knives are made in Sheffield, for although it happens to be true that knives are made in Sheffield, this is not a valid conclusion to draw from the two previous statements. (Figure 5.)
The limitation of logic is that it is reasoning without experiment. Consider the syllogism:

All drinks sold in public houses are alcoholic.
Bitter lemon is a drink sold in public houses.
Therefore bitter lemon is an alcoholic drink.

That the conclusion is valid in logic can be shown by a diagram. We consider the whole range of drinks and then draw a box around those sold in public houses.
We are told that all the drinks within the box are alcoholic, which

we represent by an identity sign. We are then told that bitter lemon is a drink sold in public houses, therefore it must be within the box. But all drinks within the box are alcoholic, therefore bitter lemon is an alcoholic drink.

```
all drinks
sold in         ≡ alcoholic
public houses
```

But although the conclusion is valid in logic, it is incorrect in fact, because one of the initial premises is wrong, as experimental investigation can show. Logic is one of the most useful tools of the problem-solver, but like all tools it needs good raw materials to make a good product.

It is often helpful to draw boxes around statements showing where they *overlap* or where one *contains* another.

A
"Steel is made in Sheffield,
Sheffield is in England,
therefore steel is made
in England."

B
"Steel is made in Sheffield,
knives are made of steel,
therefore knives are made
in Sheffield."

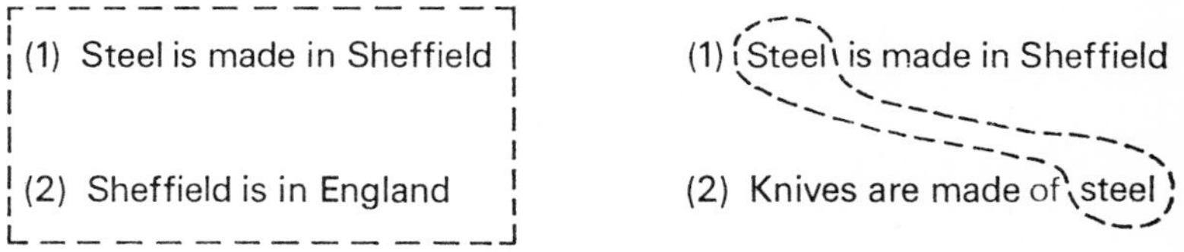

(1) Steel is made in Sheffield

(2) Sheffield is in England

(1) Steel is made in Sheffield

(2) Knives are made of steel

Statement (2) completely contains statement (1) and so the conclusion drawn is valid.

Statement (2) only overlaps with statement (1) in its reference to steel. Neither statement contains the other and so the conclusion drawn is not valid.

Figure 5 Two exercises in logic.

Intuition

Intuitive answers to problems are ones formulated by the mind, usually in a flash of thought, to which no conscious reasoning can be attached. The hunch, the inner feeling, the sudden inspiration, are examples of intuition. After the thought has arrived its correctness or otherwise may be demonstrated, but conscious rational argument arises when one is no longer struggling with the problem. Significant scientific advances have been made in armchairs, buses, beds and baths when their authors were relaxed and not thinking about their work at all. Traditionally Archimedes was grappling with the problem of Hiero's crown when he visited the baths and happened to notice the significance of the water that was displaced as he entered the bath. It is as likely that the problem was far from his conscious mind, and that while relaxing in the comfortable atmosphere of the bath he experienced an intuitive flash. Henri Poincaré (1854–1912), the great French mathematician, described one of his intuitions thus:

'Just at this time I left Caen... to go on a geologic excursion... The changes of travel made me forget my mathematical work. Having reached Coutances, we entered an omnibus to go to some place or other. At the moment when I put my foot on the step the idea came to me, without anything in my former thoughts seeming to have paved the way for it, that the transformations I had used to define the Fuchsian functions were identical with those of non-Euclidean geometry.' (R3)

Many people have had similar experiences.

Common Sense

Common sense has been described as neither common nor sense. It seems to be the ability to transfer the results of logical and intuitive reasoning from the known to the unknown. The greater the experience of an individual, the more able he is to relate old problems to new ones. The difficulty of the common sense approach is when old problems remain unsolved. Thus to people in the 19th century it was common sense that heavier-than-air machines could not fly, just as it was obvious to their fathers that iron ships would not float on the sea. It seems common sense to us that gravity shields are possible only in science fiction writing, and so nobody is engaged in research to try and make one. Yet if such could be developed it would have as big an impact on the world as did the discovery of electricity.

If worldly authority, common sense and logic are unable to explain why the sun moves daily from East to West then the way is open for intuition to suggest a supernatural reason, say that the sun-god Apollo causes it to move. If the intuitive answer comes from a person of esteem it soon becomes incontrovertible in a society where worldly authority is unquestioned.

The weakness of solving problems solely by appeal to worldly authority, or by use of logic, intuition or common sense is that there is no built-in procedure for detecting errors in truth. Surfaces are either curved or flat; the earth is a surface; therefore the earth is either curved or flat. Things slide on all curved surfaces, but do not slide on flat surfaces; things do not slide on the earth; therefore the earth is a flat surface.

If wise men and learned books state that the earth is flat; if it is common sense that people would fall off if the earth were other than flat; if one feels intuitively that the earth *must* be flat; if logical arguments can be developed to demonstrate that it is flat, then what method is available to show otherwise?

In searching for truth we must be prepared to challenge authority, to be sceptical of common sense, to suspect intuition and to question the premises of logical argument. Such an approach is found in the scientific method of solving problems.

Scientific Method

Scientific method embraces problem-solving by authority, intuition, common sense and logic. It permits any solution to a problem, however obtained, to be considered, but it rejects it unless it is in accord with the evidence. It is a method of trial and error which is scrupulously just in its trial but unyielding in its rejection of error.

The label 'scientific' has unfortunately tended to convey the impression that its use is limited to scientists, whereas in practice it is available to everyone, and as the maze in figure 6 shows is used by everyone occasionally.

Trying to find a path through a maze is a simple type of problem in which mistakes made at a junction can be corrected by returning to the junction and taking an alternative route. Suppose that in the maze we take path A at the first asterisk and path C at the second. Since this leads to a dead end we return to the second asterisk and try path D. On being frustrated again we turn further back and try path B, this time being successful. Such a sequence of events is a simple demonstration of scientific method.

The scientific method of solving a problem is to collect the facts, suggest a solution to the problem (this is usually called a hypothesis), test this solution against the facts, and if it proves wrong try another possible solution, and continue doing this until a solution arises that is successful. In the maze, at the first asterisk, the fact was that two paths seemed possible and the hypothesis that A was the right one was suggested. This led to a second problem at the second asterisk, and since neither of the two possible hypotheses there resulted in success, we returned and tried the hypothesis that B was the path. This description of scientific method is enlarged later.

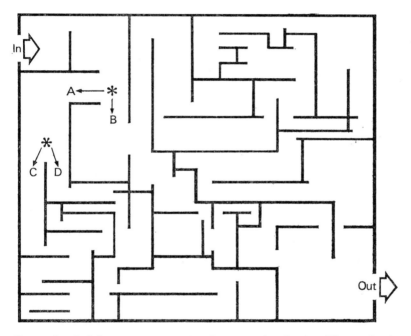

Figure 6   Solving a maze—a simple example of scientific method. Decisions taken at the asterisks are quickly reversed if they are found to lead to a dead end.

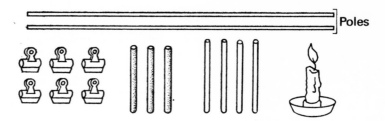

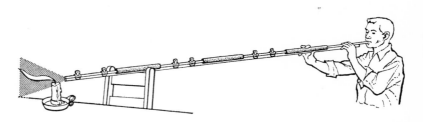

Figure 7 Maier's candle blowing experiment. Subjects had to find a method of blowing out the candle from a distance of eight feet with the apparatus shown. Subjects who had had a previous lecture on problem solving were more successful than those who had not.
Reproduced by kind permission of McGraw-Hill Book Company. From Recent Experiments in Psychology *by Crafts, Schneirla, Robinson and Gilbert.*

Can skill in problem-solving be acquired other than by years of experience of problems? There is some experimental evidence that instruction in tactics and discussion of the mental processes involved can increase the chance of successful problem-solving. Two examples are cited. In 1933 N.R.F. Maier (R4) reported the results of an experiment in which two large groups of students individually tackled a problem, which was to blow out a candle at a distance of eight feet in a limited time, with nothing more than the materials illustrated in figure 7. A group of 206 students worked at this problem, without any previous guidance from the experimenter, and 48 per cent solved it. A second group of 178 students was given a preliminary lecture lasting twenty minutes in which the following advice was discussed:

1 *Locate a difficulty and try to overcome it. If you fail, get it completely out of your mind and seek an entirely different difficulty.*
2 *Do not be a creature of habit and stay in a rut. Keep your mind open for new meanings.*
3 *The solution-pattern appears suddenly. You cannot force it. Keep your mind open for new combinations and do not waste time on unsuccessful efforts.*

Of this group 68 percent solved the problem within the time limit, an increase of 20 percent.

In an experiment lasting over a number of years M. L. Johnson Abercrombie (R5) organised discussion groups for medical students on various aspects of scientific judgment. A comparison of the results of tests taken by students who had followed the course of six 90-minute classes with others who had not, showed that the experimental group tended to discriminate better between facts and conclusions, to draw fewer false conclusions and to consider more than one solution to a problem.

The working hypothesis of the writer is that the discussions in this book will improve the reader's scientific effectiveness, or skill in problem-solving.

Chapter 2

Facts and Values

'The tragedy of science is the slaying of a beautiful hypothesis by an ugly fact.' (T. H. Huxley)

Definition of a Fact

King Charles II once asked the Royal Society why it is that putting a live goldfish into a bowl of water does not increase the weight of the bowl, but putting in a dead goldfish does. (Posing this question once at an evening lecture I was embarrassed by a widow who asserted that it must be true, because her husband had been heavier when laid out than he had been alive.) In popular parlance, King Charles had 'got his facts wrong', and so the question was not meaningful. The first essential in problem solving is to work from facts. But what exactly do we mean by a fact?

Consider the following statements:

1 I feel this thing on which I am writing.
2 This thing is a desk.
3 This desk cost me five pounds.
4 Five pounds is a lot of money.
5 This desk is a work of art.
6 This desk is really an armchair.

The first statement describes my sense experience, and, provided that my nervous system is functioning normally, is indisputable; therefore we call this a statement of fact. The second statement involves a new sense experience being compared in my brain with previous sense experiences of things which I have learned are called 'desks'. If my concept of what is a desk is generally accepted, and if my new sense experience fits in with this concept, then we may be content to call the second statement one of fact. The third statement refers to an event in which we are told I exchanged five pounds for the desk. If I can produce a receipt for this amount the statement is likely to be one of fact.

In the absence of supporting evidence the likelihood of this statement being factual may be judged on the record of my memory and of my propensity for telling the truth. I may *say* that it was a fact that I paid five pounds for the desk, X may say that it is a likely hypothesis that this is what I paid, while Y may say that it is an unlikely hypothesis. Thus statement 3 involves the exercise of judgment in deciding its verity. (It can be categorised as a historical fact. See page 35).

Statement 4 can only be described as a statement of fact if we have a definition of 'a lot of money'; the millionaire and the pauper will have different ideas of what this means. But if we choose to define 'a lot of money' as 'more than four pounds', then any normal person will agree that statement 4 is one of fact. Likewise we might decide that 5 is a factual statement if we had a definition of a 'work of art' and if the desk fitted in with this concept, but in practice we find that no adequate definition can be found. It is simple to measure a sum of money and compare it with a standard, but there is no satisfactory way of measuring an object for aesthetic appeal. Therefore statement 5 is not factual. This form of expression is usually called a value judgment, and is discussed later.

By considering reasonable definitions of 'desk' and 'armchair' we find that they are mutually exclusive and so we conclude that statement 6 is nonsense. Rather than say that 'the fact is wrong', as was said rather loosely of the goldfish problem, or 'the fact is false', it is better to say that 6 'is not a statement of fact', or, if somebody insists that it has the appearance of a statement of fact, to say that it is 'an erroneous statement of fact'. Notice that the emphasis is placed on the word *statement*. A fact is a fact, but when somebody tries to express the fact, in the form of a statement, he may fail to describe it accurately. Hence the phrase 'the fact is wrong' is nonsensical, instead it should be stressed that the description of the fact is faulty, by saying 'the factual statement is wrong'.

Strictly any factual utterance should be described as 'a statement of fact', rather than 'a fact', but usually this would be so pedantic that the latter and shorter description is used.

It is not easy to define comprehensively the word 'fact'. Consider the entry in the Shorter Oxford English Dictionary: 'Something that has really occurred or is the case; hence, a datum of experience, as distinct from conclusion.'

This is inadequate as a definition since it does not necessarily enable a fact to be clearly separated from a value. Thus if somebody says, 'It really is the case that this desk is a work of art', how do we demonstrate that this is not a statement of fact?

The first clause of the dictionary definition can be analysed into three kinds of 'somethings', namely *things, events and relationships between these*. Thus the following statements may all be about facts:

This is a desk.	(fact is a thing)
I was leaning against the desk.	(fact is an event)
The desk fell over.	(fact is an event)
My leaning against the desk caused it to fall over.	(fact is a relationship between two events)

These statements are all based on *sense data*. Others who witnessed this 'thing-events-relationship' would agree with my description of it, provided that we shared the same *language* and provided that they recognised the thing as a desk. This latter point may be expressed as: other observers must have the *stored knowledge* that the thing is a desk.

It may sound pedantic to insist that observers must have the same relevant stored knowledge about things, events or relationships in order necessarily to agree on a factual description, but figure 8 (R6) may explain why this is stipulated.

To me it is a fact that this represents a bearded man's head and shoulders. Initially the only fact apparent to the reader may be that there is an irregular distribution of black patches in a white rectangle.

Does this invalidate my use of the word fact? The reason that it does not, lies in a difference between our stores of knowledge. On scanning the area for a while, preferably with the eyes nearly closed, the reader may see the 'Hidden Man' suddenly leap into shape. Alternatively, a comparison with the drawing on page 96 may effect the transformation. Once this piece of knowledge is added to the reader's 'store' we can agree that it is a fact that a bearded man is represented. Thus it is important that a compre-

Figure 8 What are the facts about this picture? When you have read the 'stored knowledge' of the text, does your factual description change?
Reproduced by kind permission of The American Journal of Psychology.

hensive definition of 'fact' should include this requirement that observers have the relevant stored knowledge.
It is now necessary to ensure that statements such as:

It is a pity that the desk fell over.
The desk was a work of art.

are prevented by our definition from being regarded as facts.
If factual statements are limited to those to which *any observer* with the appropriate language and stored knowledge would agree, then the above statements will be excluded, since among a number of observers there can be expected to be some who will describe the desk as ugly, and others who will be unconcerned that the desk has toppled. Among a large number of observers there may be supposed to be those whose personal involvement, moral or aesthetic standards are sufficiently removed from our own to cause them to express different value judgments about the things, events or relationships under discussion.
These points enable us to advance the following *definition of a statement of fact.* A statement about a thing, an event, or a relationship is a statement of fact if a person is using his sense experiences,

stored knowledge and familiarity with the appropriate language, in such a way that any other person who receives the same experiences, has the relevant stored knowledge, and is equally familiar with the language could agree with the statement.

As mentioned earlier, it is often more convenient to refer to 'a fact' rather than to the more strictly correct 'a factual statement', and this practice will be adopted in subsequent sections.

Value Judgments

The following paragraph appeared in the correspondence columns of a British newspaper (R7) in 1964; the numbers are added for discussion purposes.

'Why do some of your correspondents take such pleasure in being snide about the "Reader's Digest"? There are some facts about that great little magazine they would do well to know. It is one of America's (1) finest (2) institutions, printing abridged (3) articles from such eminent (4) publications as "The American Legion Magazine", "True Police Cases", "The Denver Post", "This Week Magazine", "The Kiwanis Magazine", "Look", "Time", "The Saturday Evening Post", (5) and many, many others (6) of similar standing. (7) The "Digest's" timesaving (8), brilliant (9) condensations (10) of America's (11) leading best-sellers (12) are famous (13).'

The 'facts' mentioned by the correspondent have been enumerated and are now examined in terms of the above definition. In particular we consider whether other people who are familiar with the *Reader's Digest* would necessarily support his statements.

Agreement is likely on the statements that the *Reader's Digest* is American (1) and carries abridged (3) articles and condensations (10). Examination of a number of copies would provide an opportunity for agreement that these articles are drawn from the listed quoted (5) and others (6) (though there would be argument about the meaning of 'many, many'). Reliable sales statistics of American books would show whether agreement is possible on the statement that the condensations come from America's (11) leading best-sellers (12). (There could be argument about the precise meaning of 'leading' or 'best-seller'). Doubtless it would be agreed that it saves time (8) to read a condensation rather than a full-length original.

It seems then that eight of the thirteen statements are deserving of the description of fact. Yet these eight give no particular evidence to support the enthusiasm of the correspondent for the *Reader's Digest*. It is only when the words 'finest' (2), 'eminent' (4), 'similar standing' (7), 'brilliant (9), 'famous' (13) are considered, that we realise what he was excited about. But others might quarrel with these statements. Others might assess what he considers to be brilliant and famous as dull and notorious. Judgments of this type, where observers may not agree, are termed *value judgments*. Value judgments are expressions of human emotions, statements of aesthetics, of moral right or wrong, of preferences, of pleasure and displeasures. They are statements of feelings about facts.

The terms 'value' and 'value judgment' are sometimes used synonymously, but it is more accurate to distinguish between making a decision, which is the process of exercising a value judgment, and the decision itself, which is a *value*. For most purposes the distinction is trivial.

The difficulty in defining 'fact' was that a value also may be a statement based on observation, but whereas a fact will be independent of whosoever makes the observation, a value may depend on the nature of the observer.

To say that Pugwash is a humorous cartoon character is to express a value judgment, not a factual observation. Some observers will agree, others not. That this cartoon exists, that people talk of humour, that certain people make this particular statement, are facts; but the statement about its humour is a value, based on a value judgment. Likewise it may be a fact that Moses said that stealing is wrong, but the statement itself is a value. Robin Hood, for example, and the thousands of children who have read of his adventures, seemed to think that stealing is sometimes right.

Values can be divided into *moral* and *aesthetic*, and we examine each briefly.

Suppose that we had a unit of morals. The symbols +2M could indicate that a particular action was measured at two moral units, while −3M would indicate three immoral units. Actions could then be compared. If the moral measure of a man robbing a traveller was −10M, and the moral measure of his distributing the proceeds to starving children +20M, then overall we could applaud his action since the net measure would be +10M. If we had such a measure we could talk of the moral facts of the case and any observer skilled in the use of the measurement scale would arrive at the same result. But we have no such scale, and in consequence each individual works out his own, which may not only differ from that of others, but for the same individual may vary with time.

Figure 9 'This picture is a cartoon of Captain Pugwash'—a factual statement. 'Captain Pugwash is a humorous cartoon character'—a value judgment.
Reproduced by kind permission of John Ryan.

Having studied those aspects of the case which are facts (the traveller's injuries, his financial loss, the condition of the children), people make their moral value judgments. Some will judge that overall he was right, others that he was wrong.

It should be added that the decision of a jury in a court of law is quite different. A jury has to compare an action with the law of the land: facts are compared with facts. The law forbids robbery; if it is shown that a man committed robbery, then he is guilty. The jury is concerned with facts, not moral values. The statesmen who enacted the law used their moral values in deciding that this particular action was wrong, but as written in the statute book the law is a fact.

Experiences of our senses of touch, sight, hearing, smell and taste are called sense data. If our descriptions of sense data exclude our opinions, they are statements of fact. But if opinions are involved, these descriptions of sense data are likely to vary from one observer to the next and so are called value judgments; if moral considerations are not included then they can be called aesthetic value judgments. Statements to the effect that one likes historical novels, enjoys playing tennis, is enthralled by traditional jazz, but hates television parlour games and loathes unripe cheese and the feel of velvet, are all aesthetic value judgments. The word 'aesthetic' is used here with a wide interpretation such that any statement which involves the preferences of an individual and which is not concerned with moral issues can be described as an aesthetic value judgment.

The extremes of aesthetic value judgments can be understood as protective mechanisms. The human body is designed for an environmental temperature range of 60–80°F. Outside this range the body may be in danger and the individual will experience discomfort. Many poisonous substances taste nasty while many nutritionally valuable foods taste desirable. Foul tastes, bad smells, blinding lights, loud noises, painful touch sensations are all at the unpleasant extremes of aesthetic values and the reason for our displeasure can be understood in terms of evolution and natural selection. But it is not easy to explain why we can formulate aesthetic value judgments away from the extreme of discomfort. What do we mean by the joy of a painting, a sonata, a spiced dish, or a perfume? What is the nature of the delight that an idea may bring, something seemingly unconnected with the external senses?

Aesthetic values other than the extremes are in the main learned from parents and teachers. They are the product of the debate of many generations. As in morals, there are no measures in aesthetics, though somebody has suggested a unit of beauty, the milli-helen, which is that amount of beauty which will launch just one ship!

The concept of *love* is connected with moral values as the concept of *beauty* is with aesthetic values. One of the most fundamental of questions is to ask whether man searches for love and beauty of his own accord, that is by virtue of his physiology and its conditioning by social forces, or whether it is through the prompting of an external being, in other words God. An equally fundamental question asks whether man seeks *truth*, which is the concept basic to the realm of facts, because of a built-in aesthetic value judgment (I prefer the truth), or again through the encouragement of an external being. To discuss these fascinating problems and the hypotheses that have been advanced in their study, would take us too far from the objectives of this book.

It is important to realise that science cannot contribute to the discussion of aesthetic or moral problems. People who suggest that science will solve all the troubles of the world are misguided, for many problems come outside the realm of facts. Consider the problem mentioned at the outset, 'Should we eat human flesh?' First, look at the facts. Human flesh is good protein material, and from a nutritional point of view is better than the animal flesh which we eat because it contains the correct proportions of the twenty amino-acids needed to rebuild protein in ourselves. Second, look at the values. Mankind is practically unanimous in its opposition to cannibalism, and the taboos against it are the strongest of moral value judgments. The facts and the values of this problem are in conflict, and when this happens scientific method alone cannot solve the problem. Here the arguments of morality completely outweigh the arguments of fact and scientific method is powerless. It should not be concluded from this that scientists are amoral, or unconcerned with aesthetics. As much as anyone else the scientist has values; the description 'scientist' means only that in his daily work he is primarily concerned in exploring for facts, and in drawing generalisations from these.

Observed and Inferred Facts

We have discussed facts as things, events and relationships. A different classification which extends our earlier definition and which is of consequence to experimenters, is to recognise that in addition to those facts which are observed, there are facts which can be inferred without the need for observation. This can be illustrated by the story of the house with cobwebs. A class of young schoolchildren were asked what they would think if they saw cobwebs in a house. Some said they would think the housewife lazy, others that she was ill, and others that she was short-sighted. One small girl said, 'I would think that a spider had been there.' Hers was surely the only reliable conclusion to draw from the observation.

Considering this further we may say that the presence of the cobwebs was an *observed fact*, since it could be detected by sensory perception, and the comment that a spider had been there was an *inferred fact*, since it is impossible that a cobweb could form otherwise. On the other hand the suggestions that the housewife was lazy, or ill, or poor sighted, were hypotheses which attempted to explain why the cobwebs were still there. If investigation showed that the housewife was unable to see the cobwebs, but that she removed them whenever they were drawn to her attention, then we should consider the problem of why she did not normally clear them solved. What was a hypothesis would have become an inferred fact, and at the same time the other hypotheses would be rejected. We could now say that she did not remove the cobwebs because she could not see them.

Thus two categories of fact exist—observed and inferred. The latter at an early stage of investigation may be only hypotheses, but as circumstantial evidence mounts up, we come to treat them as inferred facts.

Inferred facts are those which can be inferred by logical argument from observed facts. For example it is an observed fact that a bus passes the window, it is an inferred fact that it has arrived by way of the only road. It is an observed fact that there was only one passenger downstairs and that ten got off; it is an inferred fact that nine passengers travelled upstairs. Inferred facts could be called hypotheses, but there is so little doubt about them that we are prepared to give them the hall-mark of truth and refer to them as facts. It is very unlikely that the bus arrived by way of the fields or that nine passengers were lying on the floor of the lower deck. In a nutshell, an observed fact is what we sense, an inferred fact is what we think must be, and a hypothesis is what we postulate.

Consider a laboratory example. Iodine is added to a solution in a test-tube and a deep blue colour quickly forms. Chemists would say that it is a fact that starch is present. But the only observation is that a deep blue colour forms. However to treat every such inference as a hypothesis would be excessively clumsy, and so instead, in a case where logical arguments (which here stretch over a hundred years of chemical research) lead to the one conclusion, it is convenient to describe such an inference as an inferred fact.

Historical Facts

In this year 828 there was an eclipse of the moon on Christmas morning. And the same year King Ecgbert conquered Mercia and all that was south of the Humber. This Ecgbert led his levies to Dore against the Northumbrians, where they offered him submission and peace; thereupon they parted.'

Anglo-Saxon Chronicle

Did these events really occur, and if they did are they accurately reported here?

The difficulty of the historian, compared with the natural scientist, is that the events in which he is interested are not repeatable, and the only facts at his disposal are the writings of observers who themselves may not have culled the information at first hand. The following newspaper reports (R8) of the 1956 Highland Games remind us of the difficulties of accurate reporting:

'World champion heavy-weight wrestler John Bland... took a fall. As he rose his kilt fell off. And there he stood in his trim, white football shorts.' Daily Mirror

'... his kilt came apart, revealing a pair of tartan shorts.'
Daily Express

'... he was wearing blue wrestling briefs underneath the kilt.'
Lancashire Post

'John Bland... jumped to his feet minus kilt, but wearing trews.'
Daily Sketch

'The Queen smiled broadly... Princess Margaret laughed outright.' Daily Mirror

'The Queen burst out laughing. Princess Margaret smiled.'
Daily Express

Historical facts, as available to the historian, can hardly be described as observed or inferred facts in the senses in which we have defined these terms, and since historians insist on describing events as historical facts it is convenient to regard these as a third category of facts, the reliability of which must be recognised as depending upon the accuracy of their recorders and the consensus of available reports. This is more satisfactory than the alternative of treating every event described in historical writing as a hypothesis, the truth of which can never be demonstrated.

Factual statements, value judgments and hypotheses are summarised in chart form in figure 10.

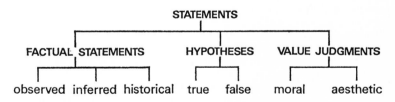

Figure 10   An analysis of statements.

Difficulties in Collecting Facts

Even where value judgments have been excluded from influencing our observations, error can creep in to interfere with the accurate recording of that which is. For example, which of the two straight lines in figure 11 is the longer? Measurement with a ruler will show that A is the longer, but the reader is likely to have decided either that B is the longer, because of the 'stretching and compressing" illusion caused by the arrowheads, or that A and B are of equal length. The latter decision would probably be based on the stored knowledge that 'I have seen these shapes before and they were the same length, even though B looked longer.'

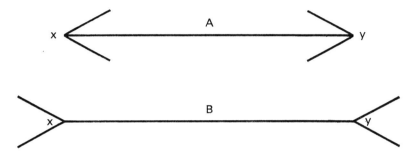

Figure 11 A well-known illusion. Which straight line (x to y) is the longer?

Stored knowledge, or previous experience, is usually of enormous importance in making observations. I could only describe the object felt at the start of this chapter as a desk because I had experienced desks before. But sometimes stored knowledge may interfere. Recently I set as a test question: 'Which is heavier, 0.99 lb expanded polystyrene or 1.01 lb water?' One candidate answered that they both weighed the same; no doubt she was familiar with the question: 'Which is heavier, 1 lb of lead or 1 lb of feathers?'

The French astronomer Lalande (1732–1807) was the leading authority in his time on the planets and their movements. On 8th May 1795 he noted a new star. Two days later he saw it in a slightly different position, which might have suggested that it must be much nearer to us than the stars. But from his experience he knew that there was no planet in this position, so he deleted the first observation in his notebook and queried the second. When Neptune was discovered in 1846 as the result of a search to explain observed perturbations in the orbit of Uranus, it was realised that Lalande had seen this planet earlier, but his stored knowledge had led him to question his observations, rather than the reverse. Ideally we must be prepared to use both observations to question our knowledge and our knowledge to question observations.

Figure 12   Examples of the influence of previous experience. Read the inscriptions aloud.

An indication of the way in which previous experience may influence observation is given by figure 12. Ask a friend to read out the list of words; he may pronounce the fourth as McHinery. Likewise the wording on the tombstone may be misread, because of the apparent familiarity of the text.

An interesting example of the influence of stored knowledge on observation is given by seeking descriptions of figure 13. The shape may be described as two faces, or as a chair leg, vase, or chess pawn etc. Yet in terms of observation it would be more accurate to describe this as a bilaterally symmetrical shape in black on a white background. The other descriptions require an extension of the shape in the mind; the visual impression has been instantly compared with a variety of profiles which have been previously

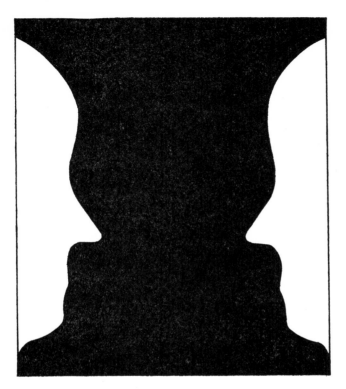

Figure 13 What is this shape?

experienced and it is identified with one or more of these. To this extent our perception has not been pure observation; a considerable element of previous experience has been linked to what is seen in order to describe it. Likewise the term 'bilaterally symmetrical shape' is drawn from previous experience. Without this ability to synthesise a description from perception and previous experience no progress could be made, but it is desirable to be ever aware of the process, in order to guard against letting it mislead. A white shape seen flitting in the night may be just moonlight reflected from a window swinging in the wind!

Observation involves perception through the senses and then synthesis with stored knowledge, and we have seen that incorrect observation may result from too great a confidence in the latter.

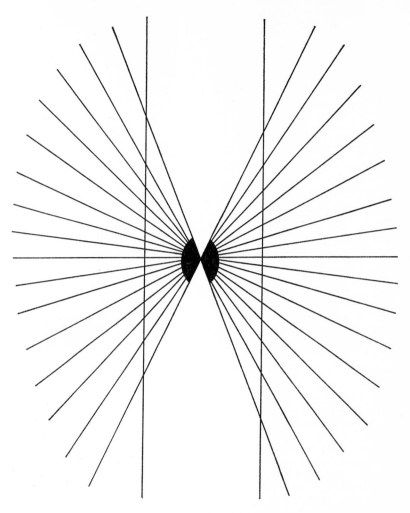

Figure 14 Mistakes in perception may be due to environmental factors. In fact, the two apparently curved lines are straight.

Mistakes in perception can also result in incorrec observation and these may be due to environmental factors, as in the optical illusion of figure 14, where two parallel straight lines appear to be curved due to surrounding lines. Also recent strong stimuli can upset the nerve cells responsible for perception. Very bright lights and very loud noises temporarily raise the threshold of perception so that an observer is partly blinded or deafened for a short while. Likewise fruit eaten after very sweet foods can taste bitter. A simple example of erroneous observation due to mistaken perception is provided by holding the left hand in hot water, the right hand in ice-cold water and then transferring them both to luke-warm water. The hands appear to observe different temperatures. What seems to be an illustration of this phenomenon was reported by Herodotus in the 5th century B.C. (R9)

'The Ammonians have another spring besides that which rises from the salt. The water of this stream is lukewarm at early dawn. At the time when the market fills it is much cooler; by noon it has grown quite cold; at this time, therefore, they water their gardens. As the afternoon advances the coldness goes off, till, about sunset, the water is once more lukewarm; still the heat increases and at midnight it boils furiously. After this time it again begins to cool, and grows less and less hot till morning comes. This spring is called "The Fountain of the Sun".'

It is frequently difficult to collect precise facts which involve people, and one reason is that the act of observing people tends to affect their actions. For example, studies of diet are difficult because many people believe that they tend to eat too much or too little, according to whether they are large or small. Consequently, if an investigator asks them to record the food eaten, or watches their choice of foods, the tendency is for them to eat what they think they ought to eat and not what they normally choose. Likewise it is difficult to measure one's own rate of breathing because the act of measuring it seems to alter the rate.

This effect of the observer on experimental data is known as the *Hawthorne effect* after a time and motion study carried out in the late 1930s at the Hawthorne factory of the Western Electric Company in the United States. The working conditions of women operatives assembling electrical equipment were varied and changes in the productivity measured. In one range of experiments the lighting in a workshop was increased and a small increase was noted in the productivity. A second increment in the lighting resulted in a further small increase in productivity. It seemed that better lighting on the job meant a higher rate of work. An obvious test of this hypothesis was to reduce the amount of lighting; the investigators expected to find a corresponding reduction in productivity, and were amazed to find instead a further slight increase. Careful study showed that production was steadily rising because the women were becoming more interested in their job, due to the activity of the investigators. Changes in the lighting had an effect that was either negligible or completely masked by other factors.

A second difficulty in collecting facts about people arises when interviews or questionnaires are used to assemble the facts about people's opinions. (Whereas the opinions themselves are value judgments, the numbers of people who think one way or another are facts). The difficulty is that often the answer given to a question depends upon the actual wording used. Some years ago a question about democracy was asked of two large matched groups of people in the United States. (R10) The first group were asked:

'Do you think that the U.S. should allow public speeches against democracy?'
and the second group
'Do you think that the U.S. should forbid public speeches against democracy?'

The replies were:

1st group		2nd group	
should allow	21%	should not forbid	39%
should not allow	62%	should forbid	46%
no opinion	17%	no opinion	15%

The two groups of people were quite comparable and so it would be expected that the range of opinions of one group on this subject would be the same as that of the other. Since 'forbid' is the antonym of 'allow' it might be expected that the percentage in the first group saying 'should not allow' would be the same as that in the second saying 'should forbid'. That this was not the case illustrates the difficulty of collecting facts about people's value judgments.

In 1964 the principal of a mixed training college in the North of England wrote to all the parents of his students to enquire whether they would object to their son or daughter entertaining students of the opposite sex in his or her *bedroom*. The parents of 261 of the 297 students gave what the principal described in an educational newspaper as 'an unequivocal "no" to bedroom visits'. Next week the newspaper carried a letter from the principal of another mixed training college in the North of England which reported that in two years only one of his students' parents had refused permission for such. In this case the principal's letter had referred to *study-bedrooms*. In both cases the students' rooms were both study and bedroom, but the different names given to it evoked such different opinions as to produce entirely opposite results. (R11)

Statistical Facts

When railway closures were under public discussion in 1963 the secretary of the Central Transport Consultative Committee sent the following letter to a national newspaper (R12):

'Sir, Your correspondent says that statistics can be made to show anything. That is one of the reasons why consultative committees do not take them into account. They only take hardship into account when reporting to the Minister on railway closures, and this is a human problem which has no need of support from statistics. Yours faithfully',

It is a popular myth that statistics can be used to prove any particular case. Consider the two graphs (R13) of figure 15 in

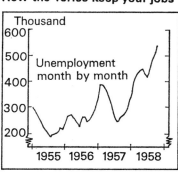 

Figure 15 Charts of actual unemployment figures which could be used as propaganda by either major political party.
Reproduced by kind permission of The Economist.

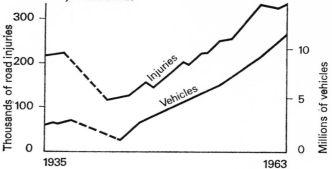

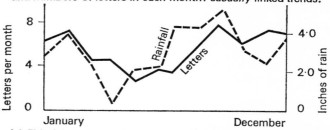

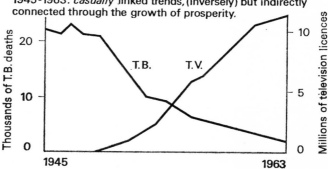

Figure 16

which the left hand graph might represent the publicity of one major political party and the right hand graph that of the other. Notice the differences in the scales. In both cases facts are being presented, taken from the *Monthly Digest of Statistics*, and if they appear to conflict the error is in the eye of the beholder. In examining these graphs it is not enough to use the simple value judgment test 'a slope left to right downhill is good and a slope uphill is bad'; it is also necessary to decide over what period of time the hill should be studied, and what degree of slope is significant.

Similar differences in the presentation of a fact can be made in the choice of words in a sentence, for example:

'He is six feet tall.'
'He is only six feet tall.'

It remains a fact that six feet is the dimension under discussion. Likewise the figures in charts and graphs are unalterable, (provided that they are facts and not fiction) although the observer may find it desirable to reorganise them into a different setting. Presentations that mislead in this way might well be termed 'intellectual illusions.' Another reason why statistical facts are sometimes denigrated is that correlations are not always understood. The three graphs of figure 16 show sets of data of which the first two can be described as showing close correlation and the third close negative correlation. (The mathematics of correlation coefficients is too complicated to discuss here.) The data are facts, and the correlations are facts, but they are not necessarily linked by cause and effect relationships. Unless other evidence is available, such relationships can only be unsubstantiated hypotheses which should be treated with scientific scepticism.

Figure 16(a) shows annual road injuries in Britain plotted with the numbers of registered vehicles, and it is reasonable to suggest that the magnitude of the latter is the cause of the magnitude of the former; in other words it is highly likely that these two sets of data are *causally* linked. Figure 16(b) shows the monthly rainfall in England and Wales in 1957 and the number of letters in each month. They are neatly correlated, but since there is no known reason why the weather should be influenced in this way, we may describe the sets of data as *casually* linked. (This assertion does not rule out the hypothesis of a causal link; it is just extremely unlikely since there are no facts which are known to support it.) Figure 16(c) shows the increase in television licences and the decrease in deaths from tuberculosis in this country. No one would claim that television cures tuberculosis, so again we have a casual link, in this case a negative one. But there is a connection between them in so far as they are both attributable to the growth in the twentieth century of scientific knowledge and technology.

Statistical facts have been abused ever since Disraeli uttered his notorious remark, 'There are lies, damned lies, and statistics.' It is the uses to which they are put rather than the figures themselves which should be viewed with suspicion.

Generalisations and Explanations, or Laws and Theories

Sometimes a group of facts is found to fit into such a pattern that a *generalisation* can be stated which accommodates them all. For example, a visitor to this country studying the numbering of houses in streets might notice that odd numbers are restricted to one side and even numbers to the other. Observing this in a large number of streets he might formulate this as a generalisation: 'In the streets which I have examined the odd and even numbered houses are always on opposite sides'. Since it is a relationship which can be verified by other people this generalisation can be described as a fact. Suppose however the investigator feels that it is reasonable to extend this generalisation to other streets which he has not seen, but which he expects to follow the same rule. He may suppose that his findings are universal and state the broader generalisation: 'In all streets the odd and even numbered houses are always on opposite sides.' This generalisation, stretching from the known to the unknown, is typical of scientific laws and can no longer be described as fact since it is fact plus hypothesis. We must always be prepared to find that the hypothesis is in error, and that the generalisation is not universally valid.

Generalisations in science are sometimes called *rules* and *principles*, like Fleming's Left-Hand Rule and Archimedes Principle, but more usually are known as *laws*. Some laws are believed to be absolutely true, e.g. the Law of Conservation of Mass-Energy, some are approximate, e.g. Boyle's Law, some are only true statistically, that is they may not apply to small samples, e.g. Mendel's Laws, and some have important exceptions, e.g. the Law of Constant Composition of Chemical Compounds.

A *theory* is an explanation which encompasses a number of facts. The visitor studying house numbers would doubtless be curious as to the reason for his generalisation and might follow up several hypotheses. Perhaps having tested several of these he might suggest that the G.P.O. has a regulation requiring the odd and even numbers to be on opposite sides of roads. This would be his theory: it would be an explanation of the observed facts.

Boyle's Law is a generalisation relating the pressure of a gas to its volume at constant temperature. The Kinetic Theory of Gases offers an explanation of this law in terms of the random movement of molecules. The generalisation that juvenile crime is increasing may lead to the theories that there is too little discipline in the home, or that the police are bringing more juvenile cases to the courts.

Notice that both laws and theories start as hypotheses, and will both be subjected to rigorous scientific testing against the known facts. The terms law or theory imply considerable confidence in their correctness, but if facts can be found which contradict them it is scientifically essential that they be modified or rejected. On the other hand sometimes the evidence for a law or a theory becomes so overwhelming that it is given the status of a fact—that is, an inferred fact.

It has been known for a very long time that the perception of colour depends upon relatively bright lights; under low illumination objects lose their colours and appear as shades of grey. This is clearly a generalisation. The Duplicity Theory of Vision advanced an explanation of this in terms of different functions and different thresholds of sensation of the rods and cones of the retina. Research eventually built up such evidence supporting this theory that today the actions of the rods and cones are considered to be facts, or more strictly inferred facts. Likewise the existence of atoms is today considered to be an inferred fact, whereas for more than a hundred years after Dalton advanced the Atomic Theory this was merely a successful explanation of the observed phenomena of chemistry.

Chapter 3

Scientific Method

'We are coming to understand that science is not a haphazard collection of manufacturing techniques carried out by a race of laboratory dwellers with acid-yellow fingers and steel-rimmed spectacles and no home life. Science, we are growing aware, is a method. . . .'

(J. Bronowski)

Intellectual Honesty

Scientific method is described in Chapter One as a problem-solving procedure in which suggested hypotheses are tested against the known facts until a hypothesis is found which fits them all. It is proper to mention that this description of scientific method is not universally accepted. Philosophers have written at such length of the Baconian inductive method and the Cartesian deductive method as components of scientific method, that some practising scientists, finding no similarity between the philosopher's writings and their own work, have denied the existence of scientific method. Other scientists however, especially those engaged in teaching, have recognised that common to all the work we have learned to call 'scientific,' is the application of intellectual honesty to problems of fact.

It is considered to be intellectually dishonest to suppress information, or to avoid looking for information which might damage a favoured hypothesis, or to refuse to abandon a hypothesis which is contradicted by reliable evidence. Coupling the concept of intellectual honesty with the active and accurate search for facts by experiment and observation leads to the method suggested in these pages. It is no argument to say that because the historian, the lawyer and the detective may use the same procedure, we cannot call it scientific; alternatively we may call them social scientists.

Simple Problems

We have seen that solving a maze provides an example of scientific method in action. A simple crossword puzzle provides a similar illustration, which although equally trivial, enables the stages of the method to be seen clearly.

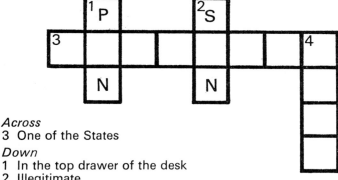

Across
3 One of the States

Down
1 In the top drawer of the desk
2 Illegitimate
4 First man

Taking 3 across, we have two facts: the clue, and the number of letters. Supposing that we assume this refers to the United States of America we may put forward the hypothesis that *Arkansas* is the answer. We test this hypothesis by trying 1 down to see if it fits. The only reasonable hypothesis for this are *Pin* or *Pen*, the alternatives of *Pan* or *Pun* seem unlikely. Either of these hypotheses rules out Arkansas, and likewise dismisses the possibilities of *Colorado*, *Illinois*, and *Maryland*. But *Delaware* would fit the facts, with 1 down being 'pen'. We test this hypothesis further by looking at 2 down. The reasonable hypotheses here seem to be *Sin* or *Son*, and so Delaware must be rejected. The hypothesis *Nebraska* likewise fails on the test of 2 down. Groping around for other States we

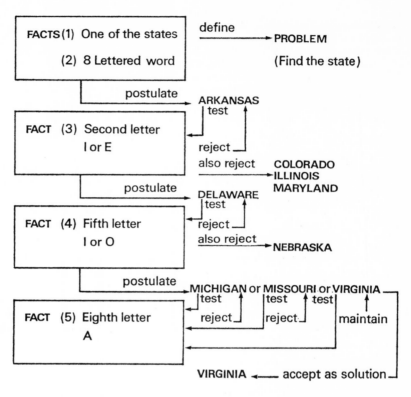

may think of *Michigan, Missouri,* or *Virginia,* each of which fits the available facts. But when we consider 4 down, only one hypothesis seems possible—*Adam*, and so on this final test we are left with the hypothesis that 'Virginia' is the answer, with 'pin' and 'sin'.

Our method has been to consider whatever eight lettered answer arrived in our mind (the order of arrival being intuitive) and to test it, and reject if necessary. If we had thought of Virginia first time, our labour would have been short, but whatever route we followed we would still have arrived at the same answer.

The facts used and hypotheses advanced and tested can be set out diagrammatically.

This crossword, and the maze of a previous chapter, are simple problems, but they form a useful introduction to scientific method by demonstrating the trial and error nature of the method, and also by showing that for simple problems it is used by everybody and is a part of what we call our common sense. We now consider a less trivial problem, which is presented in the form of a dialogue which has become familiar to me in the course of lectures.

Bassey: Please watch carefully the apparatus in my hand. (He holds a red plastic funnel which is fixed to a long wide corrugated tube, one end of which disappears under the bench. It looks like the 'snake' of a vacuum cleaner.) I put a table tennis ball in the funnel and switch the apparatus on. (He clicks a switch under the bench; the machine makes a roaring noise like a vacuum cleaner.) What do you observe?

Student: The ball is twisting jerkily at the bottom of the funnel.

Bassey: We will call this a fact, since it is based on your sensory perception. Would you suggest a hypothesis to explain why it is jumping about?

Student: I think air is blowing up the tube, but it cannot be very powerful or it would blow the ball out. I don't think it is sucking because the ball would be held tight in the hole.

Bassey: This is your hypothesis?

Student: Yes.

Bassey: Let us test this hypothesis by collecting some more facts. (He turns the funnel, still connected to the 'snake', upside down.) What do you observe?

Student: (Surprised) The ball is still there! It hasn't fallen out.

Bassey: You agree this is a fact?

Student: Yes.

Bassey: Does this affect your hypothesis?

Student: Yes. The air is being sucked into the funnel. The funnel surface must be rough causing the ball to jump about.

Bassey: You have changed your mind?

Student: Absolutely.

Bassey: You have rejected your first hypothesis and constructed a second one. Shall we test it against more facts?

Student: Is there any need?

Bassey: Watch (He removes the funnel from the 'snake', holds the 'snake' so that the opening points upwards, and places the ball six inches above the opening. The ball stays there, twisting and turning.)

Student: (Exasperated) Why does the ball stay there?

Bassey: You have the facts—make your hypothesis. The

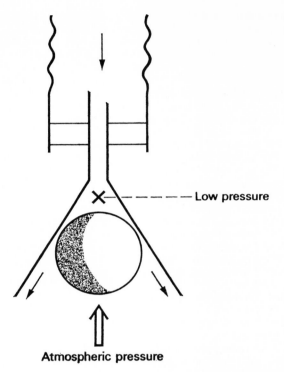

Figure 17 The ball in the funnel problem. This is an illustration of Bernoulli's Principle, that in a moving gas, where the velocity is high the pressure is low, and where the velocity is low the pressure is high. In the neck of the funnel the air from the blower is moving faster than at its mouth and so the pressure there is lower than atmospheric pressure which consequently holds the ball in place.

                apparatus is unchanged, save that I have taken the funnel off.
Student: Well, it must have been blowing all the time.
Bassey: Let us test this new hypothesis. (He replaces the funnel, puts the ball in position, turns the funnel upside down, and points it at chalk dust nearby. The dust is blown away. The ball stays in the funnel as before, jerking about.)

Student: Yes, it was blowing all the time. But I have completely reversed my opinion twice. I feel foolish.

Bassey: No. This is not foolish, it is scientific. Each of your hypotheses was reasonable on the evidence available at the time. To change your mind in the face of new facts is to be scientific. (He goes on to explain why the ball stays in place, as shown in figure 17.)

Changing the Mind

One of the distinctive features of scientific work is that changing the mind is considered laudable, if the facts so require. In some walks of life it is considered to be only a lady's privilege to change her mind, but as far as a scientist is concerned it becomes an obligation when his hypothesis is found to be at variance with the facts.

T. H. Morgan started his work on the fruit fly with a profound distrust of Mendel's laws of inheritance, but later not only changed his mind but provided an explanation of these laws in terms of the chromosomes. Had he been a politician in outlook, the history of modern genetics might have been quite different, for politicians seem unable to reverse their opinions. If the Member for A can show that the Member for B was whistling a different tune in a speech a few years back he is considered to have achieved a political victory. It is apparently improper for the Member for B to change his mind.

In 1962, chemists the world over had to change their minds on a piece of GCE A level chemistry. Since the discovery of the elements helium, neon, argon, krypton, xenon and radon by Ramsey in the years up to 1900, it had been universally taught that chemical compounds of these could not be formed. They were called the *inert*, or *noble* gases. Today the former name is obsolete, for they are known not to be inert. Two years after the secret of how to make fluorides of xenon had been published more than 100 papers on noble gas compounds had appeared in the chemical literature. Chemists were quick to change their minds once the first confirmatory experiments had been reported.

Compare this with the reception of the finding by the mathematician and Minister of the Church of Scotland, A. Q. Morton, that computer analysis of the 14 Pauline Epistles in the New Testament indicates that only five were written by St. Paul. Using counts of sentence length and frequencies of occurrence of words like 'and', 'but' and the parts of the verb 'to be', seven measures of authorship were developed which were tested and substantiated on the known works of a number of ancient and modern writers. Morton writes of his attempts to publish his discovery:

'The first technical article I wrote I sent to the "Scottish Journal of Theology". It arrived back within three days. I sent it to the "Expository Times". A letter came back, "Dear Mr. Morton, I do not understand this but I am quite sure that if I did understand it, it would be of no value". I then sent it to "Science News", whose editor came up to see me about immediate publication.' (R15)

Since the scientist is always prepared to change his mind if necessary, he tends not to make sweeping assertions, but to speak with caution and humility. That this is not always understood by others is shown by the following newspaper review of a science programme on television.

'So much did "Eye on Research" surpass itself last night that I have not the slightest doubt that the vast majority of the viewing public did not have a clue what the scientists on the screen were talking about. Furthermore, it was hinted that not all the lecturers were one hundred percent certain either. I heard the words "We believe" and "We think" far too often.' (R16)

The Scientific Method Routine

The routine procedure of scientific method can be set out in four stages:

1 Define the problem.
2 Collect facts relevant to the problem.
   a. Attempt to eliminate bias and prejudice in recording observations.
   b. Where possible repeat experiments to confirm observations.
3 Arrange and re-arrange the facts until some apparent solution to the problem arises. Call this a hypothesis.
4 Test this hypothesis against the known facts. If it is not in accord with all of them, reject it and seek another hypothesis. If it is supported by the known facts, seek further facts against which it may be tested; particularly seek facts which seem to be the most likely to test the hypothesis thoroughly. Whenever the hypothesis is contradicted, modify or reject it. Even when the hypothesis has stood the test of a wide range of facts, be

prepared to test it against fresh ones and to modify or reject it if necessary.

Diagrammatically this can be represented as follows:

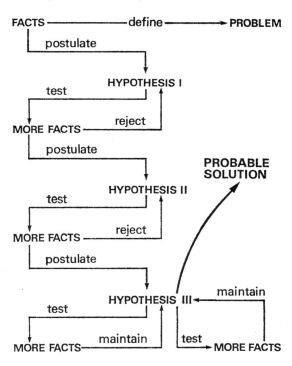

The order in which hypotheses are formulated has no bearing on the final result, though it may considerably affect the time taken to arrive at a reasonable theory. Provided each is tested thoroughly the same unambiguous answer must eventually be reached.

Stage one is the definition of the problem, but obviously this depends upon sufficient facts being available to make it clear that there is a problem. During the search for new facts it may become obvious that the problem has been ill-defined and a re-orientation may be necessary.

The word 'relevant' in stage two is all important, but elusive to define. Irrelevant facts may clutter the mind of an investigator and distract his attention from the real issues, but how does he know what is irrelevant? In the problem of the ball and vacuum cleaner, knowledge of the room temperature would hardly be expected to be relevant, but might not the weight of the ball be important? Apt judgment of what is, and what is not, relevant to a problem is one of the marks of the competent scientist. It goes without saying that it is better to include too many facts than too few. Dead wood can be pruned.

The injunction 'attempt to eliminate bias and prejudice in recording observations' is a reminder of human frailty. Suppose that a favoured hypothesis requires a meter reading to be nearer to 7 than 8. (Figure 18) It is very easy to believe that the reading shown is nearer to seven than eight.

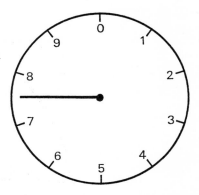

Figure 18 Beware of temptation! Suppose that if the reading is 7 this supports a favoured hypothesis; if it is 8 the hypothesis is destroyed. What is the reading?

In a lecture I once drew figure 11 (page 37) on the blackboard, making line A $18\frac{1}{4}$ in. and line B $17\frac{1}{4}$ in. By a show of hands the majority of students expressed the view that they were the same length. When I asked one student to measure them she reported that both measured $17\frac{1}{2}$ in.! She was so convinced of the correctness of her hypothesis that she misread the ruler. (I asked several students to check this before everybody was convinced.)

Bias and prejudice in the recording of observations should not be confused with the legitimate exercise of bias in the search for facts. The favouring of one particular hypothesis may indicate a line of research, and provided one is prepared to accept any unexpected

Stage 1
Eight shapes are removed from a box. Certain facts are apparent.

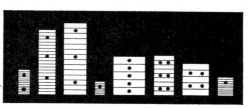

FACTS
(1) rectangular shape
(2) different dimensions.
(3) different shading
(4) different dotting

PROBLEM Is there a rule linking these properties?

Stage 2
Classify in order of height.

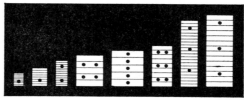

FACTS   (5) height order shows no graduation in the other properties.

Stage 3   Classify in order of number of lines per shape.

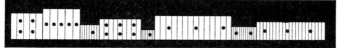

FACTS   (6) number of lines per shape order shows no graduation in the other properties.

Stage 4   Classify in order of number of lines per unit length.

FACTS   (7) number of lines per unit length order shows a uniform graduation in width of the shapes.

HYPOTHESIS
All the shapes in the box will show this same property of (7)

Figure 19   The classification of facts may lead to a generalisation as suggested by attempts to classify these eight shapes.

57

facts and to reject the hypothesis if necessary this bias may prove valuable.

A successful hypothesis goes through two phases with its inventor. In the first he tries to convince himself that it is possible and so searches for facts that are likely to support it. (His hope that it is possible can be described as bias). In the second phase he tries to show that the possible solution is the true one and so he searches for facts that will test it thoroughly, indeed to destruction if it is false. The inventor first makes his hypothesis and then strives to break it.

The processes of the invention of hypotheses are most difficult to describe. We know little about the mental genesis of ideas; of why one person is more productive than another, of why one idea may come when the inventor is struggling hard with the problem while another may come when he has abandoned the matter and is at rest. But shuffling the facts in the mind, arranging, organising, classifying them, can be significant in problem-solving. Two examples may make this clear.

Suppose we find a box of shapes. We remove eight shapes and find that they are rectangular, of different sizes and spotted and shaded differently. From these facts we formulate a problem: is there a generalisation which governs the sizes? See figure 19.

In seeking a possible generalisation we try various arrangements of the shapes. We may put them in a line in order of number of dots, or of height or width or area. Each attempt provides a fact, either of the negative result that height order shows no system, or a positive result that width order does reveal a system.

From such a fact we may advance the hypothesis that this system applies to all the shapes in the box, and we may test this by examining further shapes.

Classification in this problem is an important step towards the formulation of a possible hypothesis. It can be argued that in deciding to classify by height or width or area, we were in practice testing a hypothesis, but classification sometimes seems so remote from hypothesis formation that it is worth specifying it as a separate step in problem-solving.

Consider the Problem of Eight shown in figure 20. Eight circles are arranged in a matrix and the problem is to put the numbers 1 to 8 in the circles in such a way that adjacent integers (e.g. 3, 4 or 7, 8) are not in circles which are joined by lines.

The problem may be tackled by simple trial and error by placing integers at random, but this is a time-consuming process unless one has phenomenal luck. However if the routine discussed earlier is used the problem may be more speedily solved.

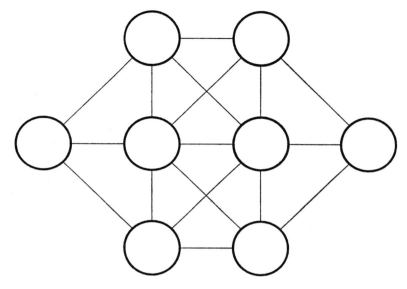

Figure 20 The problem of Eight. Arrange the numbers 1 to 8 in the circles in such a way that adjacent integers are not in joined circles.

The relevant facts are: eight circles, eight integers. There is no chance of our being misled by bias or prejudice in noting these facts, nor are we likely to gain by repeating the observation of such. (Save by reading the problem again, which is of course, always worthwhile in paper exercises.) We progress to stage 3 and begin to arrange these facts into a pattern. We find that our circles and integers can be classified:

2 circles have six links
4 circles have four links
2 circles have three links
2 integers (1 and 8) have one neighbour
6 integers (2, 3, 4, 5, 6, 7) have two neighbours

On reading this through the reader is likely to take the creative step of formulating the hypothesis: put the integers with one neighbour in the circles with six links. On testing this proposal by putting 1 and 8 in the central circles, we find that only one space is available for 2 and one space for 7. The remaining four numbers fit easily into place.

The process of classifying the circles and integers was not obviously connected with a hypothesis, but once done it provided a platform for creative thinking.

The classifications of plants and animals, by the 18th century biologists Linnaeus and Buffon, prepared the way for the theories of evolution and natural selection, while the classification of the chemical elements by Mendeleef and Lother Meyer in the mid-19th century led to the electronic theory of the atom.

The fourth stage of the scientific method routine insists that a hypothesis must be modified or rejected if facts are found which contradict it. The same is true whether the hypothesis is the simple solution to a mundane problem or if it has grown in stature to a theory or a law. Nothing is sacrosanct in science save the pursuit of truth.

Röntgen, the discoverer of X-rays, thought at one time that X-rays might penetrate to equal depths of metal specimens in which the products of thickness and density were the same. This might have resulted in an important law of physics, but on testing the hypothesis he found that it was not valid. It joined the ranks of failed generalisations.

The Phlogiston Theory of Combustion and the Caloric Theory of Heat are classic examples of theories which were widely accepted, until Lavoisier in 1777 and Rumford in 1798 demonstrated experiments which revealed facts that made them untenable.

Early in the 1939–45 war it was the belief of senior officers in the Royal Navy that in the battle between U-boats and convoys the losses of merchant ships were greater in large convoys than small ones. The simple exercise of analysing the records showed that this hypothesis was incorrect. Consequently the average size of convoys was increased from 32 to 54 vessels and as a result the losses of shipping were reduced by more than half. (R17)

It seems obvious that hypotheses should be thoroughly tested against the facts, and some people in consequence argue that this procedure is so commonplace that it is undeserving of the title 'scientific method'. Yet the history of mankind is studded with instances of unscientific thinking, or of 'failure-to-test-the-hypothesis'.

Here we consider one from the 2nd century A.D. and one from the 20th.

The father of the Greek doctor and medical writer Galen had never touched fruit and lived to be a hundred. Galen considered the two facts to be linked as cause and effect, and because his writings were widely read and acted on, millions of people over a span of 1500 years were brought to sickness and premature death from the vitamin-deficiency disease 'scurvy'.

In 1965 a committee of the U.S. House of Representatives reported that the Federal Government owned 512 lie-detectors or polygraphs for which it had paid $428,000. These lie-detectors measure physiological variables such as the electrical resistance of the skin, which changes when a subject sweats. The committee stated:

'The Federal Government has fostered the myth that a metal box can detect truth or falsehood, by spending millions of dollars on polygraph machines and salaries for hundreds of federal investigators to give thousands of polygraph examinations. Yet research completed so far has failed to prove that polygraph interrogation actually detects lies or determines guilt or innocence. . . . While federal investigators testified to their great faith in the polygraph technique, they admitted there are neither statistics nor facts to prove its value.' (R18)

Instead of rejecting a hypothesis to which contradictory facts have been found it may be possible to modify it. A number of the laws of physical science have become recognised as approximate generalisations and the originally rigid hypotheses have accordingly been modified. Boyle's Law provides an example.

Boyle's Law is one of the most widely known laws of physical science: the volume of a mass of gas at constant temperature is inversely proportional to the pressure, or, in symbols:

pv = constant (at constant temperature)

This law was first formulated by Robert Boyle in 1660.

In an experiment to verify Boyle's Law the following results may be obtained:

Pressure (atmospheres)	*Volume* (litres)	*Pressure x Volume* (atmosphere, litres)
0.10	10.0	1.0
0.50	2.0	1.0
1.0	1.0	1.0
5.0	0.20	1.0
10.0	0.10	1.0

The experimental facts are found to support the law. Suppose that we now improve our experimental technique so that pressure and volume can be measured ten times more accurately. The following results may now be obtained:

Pressures (atmospheres)	Volumes (litres)	Pressures x Volume (atmosphere, litres)
0.100	10.0	1.00
0.500	2.00	1.00
1.00	1.00	1.00
10.0	0.101	1.01

The implication of the last result would normally be that this variation from the rule was an artefact, or variation introduced by an inadequacy in our experimental procedure, rather than a failure in Boyle's Law.

Suppose that we again improve the experimental method and increase the pressure range. The following results might be obtained:

Pressure (atmospheres)	Volume (litres)	Pressure x Volume (atmosphere, litres)
0.1000	10.00	1.000
0.5000	2.000	1.000
1.000	1.000	1.000
5.000	0.2006	1.003
10.00	0.1006	1.006
50.00	0.02066	1.033
100.0	0.01064	1.064

These results are taken from some very accurate experiments with the gas hydrogen. (R19) They show that Boyle's Law is an incorrect generalisation and, therefore, following the procedure of scientific method, we should reject it. But to reject it out of hand is to overlook the point that in many engineering calculations involving pressure and volume, the result is required only to a low order of accuracy, and only rarely are pressures used which are greater than a few atmospheres. For a car tyre pressure gauge, for example, it does not matter whether the reading is 20 lb. in^{-2} or 20.1 lb. in^{-2}. A gauge based on Boyle's Law is therefore perfectly adequate for tyre pressures.

It is often preferable to state the limitations of a hypothesis rather than reject it, when some contradictory fact had been found. Thus Boyle's Law may be expressed in an absolute form as: below pressures of 10 atmospheres the pressure of a gas, under isothermal conditions, is inversely proportional to the volume to an accuracy better than 1 part in 1,000; above this pressure range the same relationship applies, but with decreasing accuracy.

Römer	1676	220,000	km/sec
Fizeau	1849	313,000	km/sec
Foucault	1850	298,000	km/sec
Cornu	1875	299,990± 200	km/sec
Michelson	1926	299,796± 4	km/sec
Rank	1965	299,792·8±0·4	km/sec

Figure 21 'Science is an infinite series of approximations to the truth' (Kelvin)' In the past three hundred years many determinations of the velocity of light in a vacuum have been made. This table shows how the measurements have become increasingly accurate, but this remains an open problem; in principle more accurate measurements could be obtained.

Most of the laws of science—physical, biological and social—have limitations to their universal applicability, but they serve the purpose of enabling predictions of phenomena to be made, provided that we are familiar with their limitations. As scientific research progresses, improvements are sometimes made in the definitions of laws so that they accord more accurately with the facts. Thus the Law of Conservation of Mass later became redefined as the Law of Conservation of Mass-Energy. The relationship between pressure and volume of a gas under isothermal conditions can be expressed in a more successful form than Boyle's Law, but being more mathematically complicated it is only applied when predictions of considerable accuracy are required.

Kelvin once described science as 'an infinite series of approximations to the truth'. As hypothesis succeeds hypothesis we approach nearer and nearer to the truth.

It should be clear from this discussion that in applying the fourth stage of the scientific method routine—a hypothesis must be modified or rejected if facts are found which contradict it—it is often more useful to modify the hypothesis than to reject it out of hand.

It may seem surprising that one scientist can sometimes passionately defend one hypothesis against another scientist who ardently holds to an alternative. For example Wegener's Theory of Continental Drift, which is represented by three maps here, has been hotly disputed among geologists ever since it was put forward in 1915. What happens is that having considered the available facts

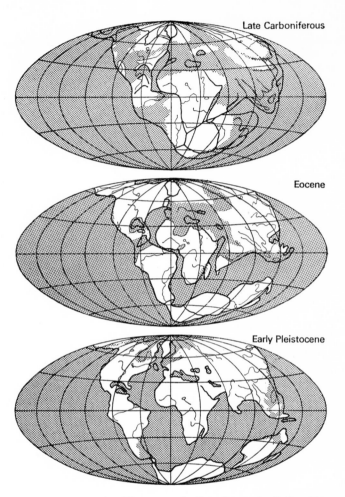

Figure 22 Wegener's Theory of Continental Drift. Wegener postulated that in Late Carboniferous times there had been one large continent, which had subsequently broken into parts which slowly drifted apart. This explained, among other facts, why the west coast of Africa and the east coast of South America have interlocking shapes. The shaded areas on the continents represent shallow seas.
After Wegener, Die Entstehung der Kontinente und Ozeane, *1915.*

and weighed up the hypotheses which have been advanced to explain them, scientists, being men and not machines, apply aesthetic value judgments and may find that they prefer one hypothesis to the others.
Whereas people must agree on facts, they may differ in value judgments, and accordingly may differ in their favoured hypotheses.

Closed and Open Problems

Black box problems offer useful demonstrations of scientific method. A simple example of a black box is a biscuit tin painted black, with an egg carton glued to the inside of the lid, corrugated paper glued to one side, and two large marbles within. The problem is to find out as much as possible about the inside without removing the lid.
This can be talked of as a model problem. It has analogies with the investigation of say, the atom, or of genes. Rutherford once likened his work to that of a man throwing stones through the windows of a room in an attempt to explore the inside. If a stone came through a window on the other side certain conclusions about the room could be drawn. Beadle likened his research on genes to the investigation of a car factory from the outside by tying the hands of particular workers as they go into the factory, and then watching to see which parts of the cars leaving the factory are missing.
The biscuit tin black box is investigated by rattling, by rolling the 'particles' and by using them to explore the inner surfaces. In lectures I have rocked the above described box to and fro and often people suggest that a cylinder is rolling across the inside. When it is apparent that the particle will also roll in the transverse direction the cylinder hypothesis is usually modified to that of a sphere. When the box is rocked upside down the particle 'disappears' in the compartments of the egg carton. As the box is turned over from this position the particle hits the walls of the tin and since two distinct hits are heard the hypothesis of one sphere is usually modified to two spheres. When the 'forbidden' action of removing the lid of the box is done, the word 'truth' is found painted on the bottom. In problem solving can we ever hope to achieve the complete truth?
In the black box investigation we could tell absolutely that the surface of the lid was different to that of the bottom. We could become positive that there were only two particles making sufficient noise to be detected, but we could not know whether, for instance, a small ball of cotton wool was also present. It would

Figure 23 A black box opened. The investigation of the black box is a model problem, where hypotheses about its contents can be tested by further observations.

require skilful X-ray photography to detect the word 'truth' written on the bottom, but it would not be possible by any presently known method to detect the colour of the lettering. Thus there are some questions that we might ask about the box which we could answer with the conviction that we knew the complete truth, there are some that we could make a reasonable attempt at, and there are others with which at present we could make no progress. It is suggested that we distinguish two types of problems of fact. Problems that can be completely solved in principle we will call *closed problems*, while those that can be in principle only partially solved we will call *open problems.*

In the experiment with the vacuum cleaner (page 51) we could pose the question 'Is the ball moving because the air is being sucked into the funnel, or because it is being blown out?' We came to an answer which was the unambiguous truth; without doubt we had completely solved that particular problem, and so we can call it a closed problem; having studied the facts we could close the discussion. There are other questions which we could only partially answer, such as 'why did the ball move about in the way observed?' Doubtless we could derive a mathematical formula to describe the motion of the ball, but this would only be approximate since in the final analysis the motion of the ball depends upon the motion of the molecules of air, about which we can have no absolute knowledge. This therefore can be called an open problem. By improving our formula we may approach nearer to the truth, but we cannot reach it completely.

The following example explains why the expression 'solved in principle' is needed. The problem 'Is the moon made entirely of green cheese?' is a closed problem because although at the time of writing no one has been able to examine any part of the moon in a laboratory, in principle this could happen. The term 'closed problem' does not mean that it has yet been solved, but that we believe it can be absolutely solved in circumstances which we can define and hope to achieve. On the other hand the problem 'of what is the moon made?' is an open problem, because the more refined the method of investigation the more information we may expect to gain. Spectroscopic analysis of light reflected from the moon's surface and telescopic study of irregularities in the surface give some clues as to composition, but a returning moon probe will provide more data. Yet this will only give information about the locality where the probe landed. The task of analysis of the whole moon, surface and beneath, is clearly beyond practicality and so the question of composition, partially answered, is likely to remain for ever open.

Further examples of open problems are: what are the properties of protons, atoms, molecules, cells, organisms, communities, planets, stars, galaxies?

As new techniques of exploration are developed, as more workers advance daring hypotheses and exhaustively test them, we slowly move nearer to the truth. But because we can never tell what further discoveries about these things may be made, the investigations of them remain open problems.

Closed problems usually involve a particular thing or event, for example, 'Who committed the murder?', whereas open problems are more usually about generalisations, such as 'Why are murders committed?' Historical problems start as closed problems but may become open if no one studies the evidence when it is available. For example, 'Who murdered the Princes in the Tower?' obviously had a definite solution, but now must be treated as an open question, since the murderer can only be defined as one of a handful of people.

The word *proved* is frequently misused. The suggested solutions to closed problems may be proved or disproved, but those to open problems may only be disproved. This is so because if a solution to an open problem were proved, the problem would become completely solved, and thereby a closed problem. The open problems of science advance by disproof. We cannot prove that a hypothesis in an open problem is right. We can only show, or fail to show, that it is wrong. Our hypothesis may be advanced in status to a law or to a theory because we fail to disprove it. Over and over again we may verify that it applies in a particular case, but this does not mean that we have proved it for all cases. The examiner who asks the student to 'prove Ohm's Law' is seeking the impossible: the boy can only verify the law in specific cases.

Mention may also be made of *phantom problems*, a term used by the physicist Max Planck. People sometimes ask 'Are there any forms of radiation which we are unable to detect?' or 'Surely space must end somewhere?' But if we cannot detect, we cannot tell whether something exists. If we were to find an 'end', we would ask 'What is beyond?' These problems only appear to be problems of fact, in reality they are problems without facts. Without facts we cannot use scientific method, and so we call them 'phantom problems'.

Chapter 4

Some Contemporary Social Problems

'One of the ways in which we may hope to solve the political, social and economic problems that confront us is by reforming our minds; and to examine these problems in the same critical, disinterested and unprejudiced attitude in which scientific men have carried out their labours and researches and reported the results of them to the world.'

(R. W. Jepson *Clear Thinking* 1936)

Throughout this book it has been stressed that scientific method is as applicable outside the laboratory as inside. In this final chapter we examine several contemporary social problems in sufficient detail to illustrate some of the ideas developed earlier.

1 Cigarette smoking and lung cancer

On Ash Wednesday 1962 the Royal College of Physicians published a report (R20) entitled *Smoking and Health*, which summarised much of the current medical discussion on the likely relationship between cigarette smoking and cancer of the lung. The following extracts are quoted from the summary.

'After its introduction to Europe in the 16th century, tobacco smoking, mostly in pipes, rapidly became popular. It has always had its advocates and opponents, but only recently has scientific study produced valid evidence of its ill-effects upon health. Cigarettes have largely replaced other forms of smoking in the past seventy years, during which time tobacco consumption has steadily increased. It is still increasing. Women hardly ever smoked before 1920: since then they have smoked steadily increasing numbers of cigarettes.

Three-quarters of the men and half of the women in Britain smoke. Men smoke more heavily than women. Smoking is now widespread among schoolchildren, especially boys.

Tobacco smoke is complex in composition. Its most important components are: nicotine, which acts on the heart, blood vessels, digestive tract, kidneys and nervous system; minute amounts of various substances which can produce cancer; and irritants which chiefly affect the bronchial tubes. The amounts of carbon monoxide and arsenic in the smoke are probably too small to be harmful. There has been a great increase in deaths from lung cancer in many countries during the past 45 years. Some of this increase may be due to better diagnosis, but much of it is due to real increase in incidence. Men are much more often affected than women.

Many comparisons have been made in different countries between the smoking habits of patients with lung cancer and those of patients of the same age and sex with other diseases. All have shown that more lung cancer patients are smokers, and more of them heavy smokers than are the controls. The association between smoking and lung cancer has been confirmed by prospective studies in which the smoking habits of large numbers of men have been recorded and their deaths from various diseases observed subsequently. All these studies have shown that death rates from lung cancer increase steeply with increasing consumption of cigarettes. Heavy cigarette smokers may have thirty times the death rate of non-smokers. They have also shown that cigarette smokers are much more affected than pipe or cigar smokers and that those who had given up smoking at the start of the surveys had lower death rates than those who had continued to smoke.'

The starting point of this discussion was the fact of the increase in deaths from lung cancer shown by the reports of the Registrar General. This single fact led directly to the question 'Why?' and the possibility of cigarette smoke being responsible was an obvious hypothesis since it is one of the few things coming in direct contact with the lungs. The comparison between the increase in cigarette smoking and the increase in lung cancer was striking (figure 24) but as Lord Clitheroe said in the House of Lords (16.7.63) (R21) in reply to a Government spokesman:

'My Lords, does the noble Lord (who had quoted results similar to figure 24) know that a great many other things have increased in exactly the same proportion during those years—betting, crime and all sorts of other things—and does he understand that statistics really cannot be used in this way.'

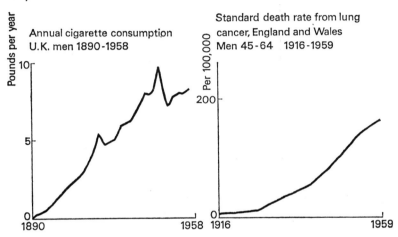

Figure 24   The increases in lung cancer and in cigarette smoking in men (R20).

Similar comparisons of the increase in cigarette smoking and lung cancer were found in all the countries studied, but this did not alter the argument that similar increases could be shown in hundreds of social variables of the 20th century.

Facts which showed that lung cancer and cigarette smoking were closely associated were published in 1956–59 by Doll and Hill in this country and Hammond and Horn, and Dorn in the United States. These workers showed conclusively (figure 25) that the larger the number of cigarettes smoked daily the greater the chance of dying of lung cancer. There can be little doubt that there is a relationship between cigarette smoking and lung cancer. However it does not follow necessarily that this is a relationship of

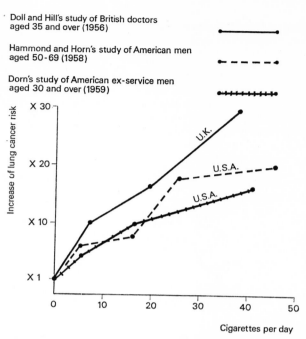

Figure 25 Three studies of the relationship between numbers of cigarettes smoked per day and lung cancer death rates. The graphs show how much the risk of getting lung cancer is multiplied in those who smoke various numbers of cigarettes per day compared with the risk of non-smokers.

cigarette smoking causing lung cancer. One of the difficulties is that the relationship is not a one to one connection. Of people smoking more than 20 cigarettes a day only 1 in 8 die from lung cancer (1956 figures); some other factor or factors at present unknown must also be responsible. Three possible hypotheses which have been advanced to explain the relationship are shown in figure 26.

In hypothesis A it is supposed that lung cancer is the cause of a desire to smoke cigarettes rather than the reverse. This seems very improbable since it implies that the process can precede the clinical symptoms of the disease by as much as forty to fifty years and requires a desire to smoke cigarettes in proportion to its liability to mature into a cancer.

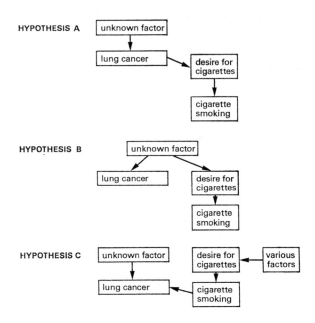

Figure 26 Three hypotheses attempting to explain the relationship between cigarette smoking and lung cancer. An arrowed line indicates a connection from cause to effect.

In hypothesis B it is suggested that there is some unknown common factor operating in some people which promotes lung cancer and the desire to smoke cigarettes. The facts of figure 25 are accommodated by supposing that the greater this common factor is, the more likely it is to lead to lung cancer and the greater is the desire for cigarettes. This hypothesis is unlikely because it has been shown that where smokers stop cigarette smoking, the chance of their developing lung cancer is reduced. Smokers who cease this habit usually do so not because the desire has ceased but as a result of a conscious decision, which frequently requires considerable willpower to enforce.

Hypothesis C fits the available facts, only a few of the more important of which have been outlined here. Notice that in common with the other hypotheses other unknown factors must also contribute towards lung cancer in order to explain the lung cancer mortality of only one in every eight heavy smokers.

The extraordinary feature of this discovery that a prevalent habit leads to a one in eight chance of death from a feared disease, is that it hardly affected the number of cigarettes smoked. Discussion of the Royal College of Physician's report was widely featured in newspapers, television and radio, but the only noticeable change was a drop in cigarette consumption of 4 percent in 1962 and a swing towards tipped cigarettes. (Figure 27.)

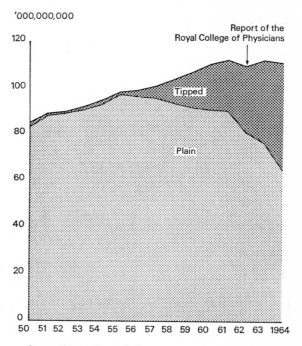

Figure 27

The cheaper price of tipped cigarettes (they contain less tobacco) no doubt contributed to their increased popularity, but the view developed that they are less dangerous than ordinary cigarettes, and their widespread advertising, although in no way suggesting this view, probably encouraged it by never denying it. The belief that tipped cigarettes afford the smoker protection is an example of a widely held but completely untested hypothesis.

On the other hand many smokers refused to accept the fact that there is a relationship between cigarette smoking and lung cancer. The aesthetic value judgment, 'I like cigarettes,' obliterated the factual evidence.

The following extracts from letters to The Guardian (R22) by a Professor of Medical Statistics and by a Member of Parliament illustrate what a third correspondent called 'a tragic and dangerous example of the gap which exists between scientific opinion and the ill-informed.'

From Professor A . . . (3rd April 1962)
'Sir, Mr. L . . ., M.P., is reported in your issue of March 27th as saying that he regards the report of the Royal College of Physicians on smoking and health as "so much unscientific tosh". On almost all matters of scientific research there is room for differences of opinion, and the report scrupulously describes differing views which have been held about the effects of smoking on health. Even more are views likely to differ on the political actions that flow from scientific discoveries.

But this report contains a carefully documented review of the available scientific evidence, prepared by some of the best medical scientists in the country. To dismiss it as "unscientific tosh" is to reveal either an alarming degree of irresponsibility or an appalling ignorance of what science is about.

Yours faithfully.'

From Mr. L . . ., M.P. (11th April 1962)
'Sir, I refuse to plead guilty to Professor A . . .'s charge of being either irresponsible or ignorant ("Guardian," April 3rd). "Tosh," I agree, is an arguable word; but I am on incontrovertible ground in saying that to select certain facts which suit a preconceived theory while blithely ignoring all those that do not is, by definition, "unscientific." The Royal College of Physicians' report contains many examples: the striking difference between cancer figures in town and country; the astounding success in causing lung cancer in mice by almost every kind of fumes except tobacco; the invented statistics of a number of countries, notably South Africa. . . . The statistics, like most statistics, can be totted

up to prove anything you want them to prove; and until their proof is a lot more convincing I for one am not going to be denied the pleasure of smoking, so richly enjoyed by all my long-lived relatives.

Yours etc.,'

From Professor A . . . (18th April 1962)
'Sir, Of course Mr. L . . . (April 11th) is "on incontrovertible ground in saying that to select certain facts which suit a preconceived theory while blithely ignoring those that do not is . . . unscientific." I believe him to be wrong in applying this stricture to the Royal College of Physicians' report. Is he really suggesting that the authors are responsible for "invented statistics of a number of countries, notably South Africa"? If so, he should give us chapter and verse. Which scientists have achieved "astonishing success in causing lung cancer in mice by almost every kind of fumes except tobacco," and why do they not publish their results?

The difference between urban and rural rates of cancer incidence is discussed at some length in the report, and no attempt is made to deny that air pollution may be one causative factor, although the evidence suggests that it is less important than cigarette smoking. The report says: "It is important to recognise that the hypothesis is not that cigarette smoking is the only cause of lung cancer."

Mr. L . . . is naturally impressed by the longevity of his tobacco-smoking relatives. Many other heavy smokers have not been so fortunate. One just cannot generalise from a handful of instances. Many factors affect the length of life and the effect of any single factor can be measured only by large numbers of observations. Whatever view one takes about the causative effect of cigarette smoking it is, I think, common ground that heavy smokers suffer higher death rates than non-smokers. But no one suggests that all heavy smokers will die early, or that all non-smokers will enjoy long lives. Incidentally, if Mr. L . . . really thinks that "the statistics can be totted up to prove anything you want them to prove," perhaps he will demonstrate to us that heavy smoking reduces the risk of lung cancer.

Yours faithfully.'

Two months after this debate the Member of Parliament was knighted in the Birthday Honours!

2 Dental caries and the fluoridation of water supplies

Dental decay is an important health problem. In a survey of 12-year-old children carried out by the Ministry of Health in 1958 the average child had 6 decayed teeth. For 19-year-old recruits to the Royal Navy at the same time the average was 16 decayed teeth. Evidence from mediaeval graveyards suggests that our ancestors enjoyed better dental health than we do, and it is believed that the recent decline is associated with dietary changes, especially among processed carbohydrate foods.

The evidence (R24) from the remote Atlantic island of Tristan da Cunha is revealing. Medical officers from Royal Navy ships visiting the island in 1932, 1937, 1952 and 1955 inspected the teeth of all the inhabitants (162 in 1932). The percentages of the population without any sign of dental caries on each occasion are shown in figure 28. The only changes that can be considered significant as far as teeth are concerned were the import of toothbrushes and large quantities of hitherto unavailable white

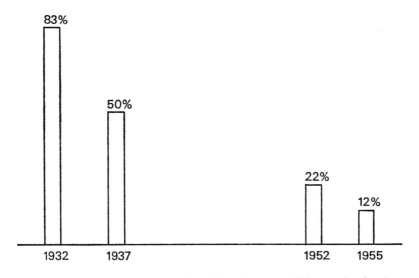

Figure 28 The decline of dental hygiene on Tristan da Cunha. The figures show the percentages of the population without dental caries from 1932 to 1955. White flour and sugar were imported from 1933 onwards—and toothbrushes!

flour and sugar from 1933 onwards. Presumably the former were not responsible!

In the late 1930's it was recognised in the United States that the incidence of dental caries in the population seemed to be inversely related to the amount of fluoride found naturally in the water supply. This may vary from negligible proportions to as much as 6 parts per million (6 p.p.m.). The higher proportions cause mottling of the teeth which, although not harmful, may appear undesirable.

Many studies supported this relationship. In Boulder, U.S.A., where there is no fluoride in the water, people in their early forties had lost an average of 15 teeth, while in Colorado Springs, U.S.A., with 2.5 p.p.m. of fluoride, the average loss of the same age group was 3 teeth. The results of a recent British survey of children aged 12 to 14 are shown in figure 29; those living in areas where there is more fluoride in the water have fewer decayed teeth.

Sufficient individual facts have been collected to establish confidence in this generalisation.

It was an obvious follow up to these studies for the dental profession to suggest that fluoride should be added to the drinking water of those areas which have a negligible amount present, and this has been done in a few parts of the United States and this country.

The experimental fluoridation of water supplies initiated by the Ministry of Health in three areas of the country in 1956 must be one of the most thorough medical experiments on a large population ever carried out. Because fluoride is known to be toxic if taken in excess it was necessary to ensure that the experiment was as safe as possible. The extensive published researches on the subject were reviewed and comparisons made of vital statistics in the United Kingdom for areas of high and low fluoride concentration in drinking water. A conference of experts agreed that 'despite considerable interest and research there is no definite evidence that the continued consumption of fluoride in water at a level of about 1 p.p.m. in drinking water is in any way harmful to health and that if any untoward effect is revealed by future research it is most unlikely to be serious.'

The three areas chosen for study and their controls are described in the following table.

experimental area	control area	regional character	water supply
Gwalchmai zone of Anglesea	Bodafon zone of Anglesea	agricultural	soft
Watford in Hertfordshire	Sutton in Surrey	residential	hard but softened before distribution
Kilmarnock in Ayrshire	Ayr in Ayrshire	industrial	soft

The local authorities in these areas had previously expressed interest in fluoridation and accepted the Minister of Health's invitation to take part in the experiment.

A standard procedure for dental examinations was devised. The following table gives results for 4 year old children in 1956, the year just before fluoridation, and 1961, when the experiment had lasted five years. The 1961 group of children had lived on fluoridated water all their lives, including the months of pre-natal development.

Percentage of 4 year old children free from caries.

	1956 (no fluoride)	1961 (after 5 years of fluoride in experimental areas— none in the control areas)
Anglesea		
experimental area	19%	43% (fluoride)
control area	15%	20% (no fluoride)
Watford and Sutton		
experimental area	35%	53% (fluoride)
control area	34%	49% (no fluoride)
Kilmarnock and Ayr		
experimental area	13%	30% (fluoride)
control area	13%	11% (no fluoride)

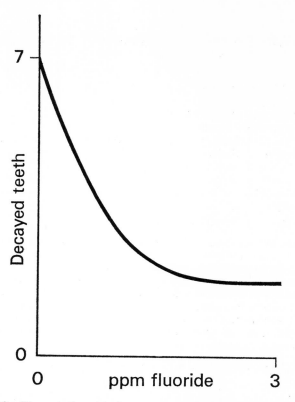

Figure 29 The relationship between tooth decay of 12 to 14 year olds and fluoride concentration in drinking water in the United Kingdom (25).

These interim results of the experiment leave no doubt that fluoridation has brought about a substantial improvement in the dental condition of most of these children.

Notice the importance of the controls in this experiment. It could have been the case that the improvement in the experimental areas between 1956 and 1961 was due to changes in diet or oral hygiene and quite unrelated to the addition of fluoride to the water. But in this event a similar change would be expected in the control areas. Indeed in Anglesea and more so in Watford and Sutton an improvement was noticed in the control areas, but less than in the experimental areas.

Apart from the thorough search for possible dangers before the experiment began, medical vigilance was maintained by the 89 general practitioners in the experiment areas being repeatedly questioned about possible ill effects to any of their patients. In five years only one possibility was suggested and this was not substantiated on investigation.

It is instructive to examine some of the opposition (R26) that has developed towards the fluoridation of water supplies. There are three main arguments:

1. *Fluorides are poisons (especially rat poisons).*
2. *Fluoridation is a political issue in which the rights of the individual (to drink or not to drink fluoridated water) are suppressed.*
3. *Fluoridation is experimental and as such no one can be sure that there will not be dangerous consequences even if none are known at present.*

The first precept for problem solvers is: recognise which of the arguments are facts, which are value judgments and which are hypotheses. The first statement is not a fact as it stands; whether something is a poison or not depends entirely on the dose involved. Strychnine is both a fatal poison and a medicine according to the amount taken. Even common salt, sodium chloride, can be poisonous if taken in massive amounts. Certainly sodium fluoride is used as a rat poison, but in a very much larger concentration than 1 p.p.m. Since there is no known evidence that fluorides at the concentration used in water supplies are harmful in any way, this statement must remain as a hypothesis. The question 'Is fluoride harmful?' is an open one. However much we know, we cannot be sure that an example of injury will not be found. Hence this hypothesis is legitimate, but deserves to carry no influence until supporting facts are found.

The second statement is based on value judgments and is therefore a matter on which a range of opinions may be held. It is interesting that a leaflet published by the London Anti-Fluoridation Campaign in 1965 argued from this position alone and denied that fluoridation was a medical/dental issue. The merit of such an argument is that no-one can disprove it!

The third statement is true of every new venture attempted by mankind. To halt action on such grounds would be to stagnate. In essence it is a hypothesis, referred to above, resulting from the open problem 'Is fluoride harmful?'

Mausner (R27) (1955) made a detailed study of public reactions to the proposal to introduce fluoridation in the town of Northampton, Massachusetts, U.S.A. A questionnaire investigation, the results of which were analysed in terms of those for and those against fluoridation demonstrated the extent of bias on what seem to be straightforward questions. Two of the questions asked, and the answers obtained, are given below.

percentage replies to:		Attitude to Fluoridation	
		For %	Against %
'Scientific bodies like the American Dental Association and the U.S. Public Health Service are the best sources for facts about fluoridation.'	Agree	92	50
	Disagree	2	35
	No opinion	6	15
		100	100
'Fluoridation has been a success wherever it has been tried.'	Agree	45	6
	Disagree	6	57
	No opinion	49	37
		100	100

In this town, when fluoridation started, a petition demanding that it be stopped was signed by 10 percent of the electorate. Consequently, a local judge issued an injunction which banned the addition of all chemicals to the water. As a result the water commissioners halted chlorination as well as fluoridation and then issued a radio appeal to the public to boil drinking water! Eventually a referendum was held and fluoridation was defeated by two votes to every one in favour.

3 The Birmingham dipped headlights experiment (R29)

For a fortnight in March 1962, motorists in the City of Birmingham were encouraged to drive with their dipped headlights on after dark, instead of the usual sidelights only. It was believed that if motorists could be persuaded to do this the accident rate at night time would be reduced. A large proportion of motorists co-operated and there was a fall in the accident rate. Moreover an opinion poll of road users showed that the use of dipped headlights by moving vehicles in built up areas at night was acceptable. In consequence a

five month experiment was planned for the following winter, to last from 1st November, 1962 until 31st March, 1963.
The Lord Mayor of Birmingham sent a report to the Minister of Transport in May 1963 from which the following extracts are quoted:

'The experiment, which received a large measure of publicity, resulted in a significant fall in the accident and casualty rates at night. Co-operation from motorists was approximately 70% on badly lit streets and 45% on well lit streets during the campaign.'
These figures compared with 12% and 5% driving on dipped headlights before.

'During the experiment there were, at night:
138 fewer accidents	(-16.9%)
19 fewer fatalities	(-49.0%)
84 fewer injuries	(-8.5%)
16 fewer pedestrians killed	(-55.0%)
28 fewer pedestrians injured	(-9.2%)

compared with the same period 1961/2. There is no doubt that a large proportion of these reductions were made through the use of dipped headlights.
The experiment had the effect of substantially cutting accident and casualty rates.'

Here we have a good example of the way that a civilised community should run its affairs. Instead of trying to enforce a possible method of reducing road accidents it was first tried out in a large city for a five month period. The hypothesis was tested; and to the delight of its initiators it seemed to be valid. The Lord Mayor wrote to the Minister of Transport:
'I am confident in the belief that the use of dipped headlights on a national scale would show a substantial reduction in road accidents at night.'
But was the hypothesis adequately tested? The experiment can be described thus:

Hypothesis: *That the use of dipped headlights at night will result in less accidents.*
Fact one: *Reduction in accidents compared to previous year.*
Fact two: *Increase in use of dipped headlights compared to previous year.*

Were these the only significant differences noted between the two years? Are there no other relevant facts?

In August 1963 the Road Research Laboratory published a detailed study of the experiment, in which it was shown that at least three other facts were significant. Considerable improvements in street-lighting had been made on a number of roads, radar speedmeter checks by the police had started shortly after the dipped headlights campaign, and the weather had been much more severe than the previous winter. Any or all of these could be responsible for the improvement observed, and so could the resultant extensive publicity about road safety.

In order to evaluate the results of the experiment unambiguously it was necessary to eliminate these three possible causes. Changes in the street-lighting were excluded by omitting the 14% of accidents that had occurred in the relevant streets from the analysis of both the test year and the control. Although the radar speedmeter was not used at night, motorists had no reason to be aware of this and so it was assumed that the deterrent effect of the speedmeter on speeding, and therefore on accidents, would be equally effective day and night. Likewise it was assumed that the effects of bad weather in reducing vehicle speeds and limiting pedestrian travel would be equal day and night. Consequently it was argued that if a significant difference could be found between the day-time and the night-time accidents in 1961–2 and 1962–3, this could probably be attributed to the dipped headlights campaign. Dipped headlights could only affect the night-time accidents. The relevant statistics are shown in this table:

Injury accidents in Birmingham (excluding roads where there were light improvements)

	Light	*Dark*	*% in dark*
November 1961—February 1961	504	592	54%
November 1962—February 1963	425	482	53%
Percentage reduction in 1962/3	16%	19%	

When only a slight difference exists between two figures, as here between 16 percent and 19 percent it is difficult to decide whether this has arisen by chance or due to some slightly effective cause. A rather complicated statistical procedure exists which suggests whether it is more or less probable by a one in twenty chance that there is some effect being measured. The result in this case was that if there is an effect it cannot be detected. More detailed analysis of the results, in which the accidents in badly-lit streets were separated showed that about one third of the night accidents occurred in the badly-lit street and here the use of dipped headlights was probably effective.

It was of interest to show whether the weather or the increased local interest in road safety, catalysed by the publicity for the dipped headlights campaign and the police speed traps, was mainly responsible for the sharp decline in road accidents of both day and night. For this purpose three towns were chosen which were sufficiently near to Birmingham to have experienced similar weather and in which there had been no recent extensive road safety campaigns, or major road works, or radar speedmeters used. Leeds, Nottingham and Sheffield served as these control towns, and the accident position in these is shown in this table.

	Light	*Dark*	*% in dark*
November 1961—February 1962	1 014	1 036	51%
November 1962—February 1963	951	1 001	51%
Percentage reduction in 1962/3	6%	3%	

The only difference likely to be important between the two years here is the weather. Notice that the reduction is considerably less than that observed in Birmingham. The Road Research Laboratory Report suggests that the order of importance of the three possible factors affecting the Birmingham night accident rate was road safety publicity, the weather, and lastly, dipped headlights.

This illustrates the difficulty in obtaining unambiguous results from a social experiment. In order to obtain a clear result which demonstrates that one particular change is responsible for an observed effect, it is desirable that no other changes occur at the same time. This is often impossible to achieve.

4 Learning to read by the Initial Teaching Alphabet

Our educational system is riddled with untested hypotheses. This is due not only to the prevalent view that practices which are hallowed by time must be the best possible ('otherwise they would not have lasted'), but also to the difficulty of carrying out adequate experiments. One of the biggest obstacles encountered in experiments with new teaching methods is the Hawthorne Effect (see page 41). In experiments on teaching, any improvement in learning may be due to interest aroused in both teachers and pupils by the experiment itself, rather than by the subject matter.

ţhe
iniʃhal teeçhiŋ
alfabet

caracter	sampl-wurd	caracter	sampl-wurd
æ	æmiabl	y	yeesty
b	brambl	z	zeeroe
c	caustic	ȥ	mœȥeȥ
d	dreȥden	wh	whot's whot
ee	eevenly	çh	çhurçhill
f	fanfær	ţh	ţhisl
g	gregory	ɉh	ɉhis'll
h	hω's hω	ʃh	ʃheepiʃh
ie	iesicl	3	televiʒon
j	jinjer	ŋ	iŋkliŋ
k	kiŋ-cup	r	erţh-wurm
l	lateral	α	faţher
m	memœ	au	auniŋ
n	nueȥ-lien	a	averæj
œ	œverţhrœ	e	emperor
p	peepl	i	iŋgliʃh
r	rœtor	o	obloŋ
s	scriptuerȥ	u	upţhrust
t	test-matçh	ω	gωd-lωkiŋ
ue	uenion rωl	ω	ωziŋ
v	vœtiv	ou	outiŋ
w	willœy	oi	oiliŋ

Figure 30 The characters of the initial teaching alphabet.

Any new method, good or bad, may generate temporary enthusiasm, which may quite obscure any improvement in learning that is due to the new method. Indeed one cynical writer has said:

'In educational experiments, no matter what the hypothesis under test, the experimental classes do better than the control classes.'

The experimental study of the Initial Teaching Alphabet (i.t.a.) provides an example of the difficulties.

This alphabet, designed by Sir James Pitman, consists of 44 characters, the additional ones being for those phonemes (speech sounds) which have no single letter in traditional orthography. Certain irregularities in familiar spellings are removed and capital letters are merely enlarged lower case letters. The object of using the alphabet is that children should learn to read with these characters and later transfer to traditional orthography. The basic hypothesis is that learning to read by this method speeds up the process of acquiring skill in fluent reading.

By February, 1965 some 2 000 children were learning to read with i.t.a., in an experiment organised by a Reading Research Unit established in the University of London, and described as follows:

'Each school acts as its own control: objective tests of attainment in reading and other subjects in the experimental classes are compared with results from the same school's control classes of children who have their reading instruction in conventional English orthography. Also there is a matched control school sample where i.t.a. is not used at all. These control schools have been provided with refresher courses and their teachers attend regular meetings to discuss reading in order to match the Hawthorne effect which may be generated by the training and research meetings of the i.t.a. teachers.

A number of criteria are used to measure the effects of i.t.a. Parallel records of progress are kept by i.t.a. and control classes, and both groups of children undergo regular testing with rigorously objective tests of relevant attainments. This testing programme is designed to provide data both on the beginning reading stage with i.t.a. and on the later stages after the transfer to reading in the conventional alphabet and spelling.

Although the interim results have been encouraging, it is still too early to give any definite judgments or firm conclusions (R29).

Unfortunately the attempts to balance the Hawthorne effect by arranging meetings and courses for the teachers in the control classes do not appear to have been very successful (R30). The experimental classes had new books and apparatus, written in the i.t.a. orthography, and lectures, meetings and a bulletin were available for their teachers. The control classes used the traditional books and although some meetings were arranged for their teachers it appears that these were not successful. Considerable publicity in the press and on television for the experiment was attracted and many visitors attended the experimental classes. There was no publicity and few visitors for the control classes.

It had been intended that the only difference between the experimental and control classes would be the change in orthography. But it is not easy to see how the teachers of the experimental classes could be prevented from becoming more enthusiastic about teaching reading than their colleagues with the control classes. Consequently it is likely that two differences developed between the experimental and control classes—a change in orthography and a 'drive' on reading. Better reading results would be expected whatever method of teaching is used, when teachers concentrate their energies on this particular part of the school curriculum.

At the time of writing this experiment is still in progress, but it is questionable whether the experimenters will be able to distinguish between the effect of the new alphabet and the effect of the emphasis on learning to read created by interest in the experiment. Again this illustrates the difficulty of social experimenting.

5 Crime and punishment

A headmaster inexperienced in television watching complained to a daily newspaper (R31) in 1961 about an evening's viewing.

'In three separate programmes during the evening one person threatened to kill another, with considerable venom, emphasis and reiteration. Such remarks as "Do you kill merely for the pleasure of it?" and "I'll kill you when I grow up" (a small child to his father) recur to mind. I presume that this is a fair sample of what goes on, dreary night after night, as the moral values of a sickening civilisation swiftly disintegrate (as swiftly at least as the increase in crime shows) . . .'

and a Member of Parliament wrote (R32):

'Is the rise in the crime rate really 'startling'? It is true that television strengthens the tendency to crime by making it

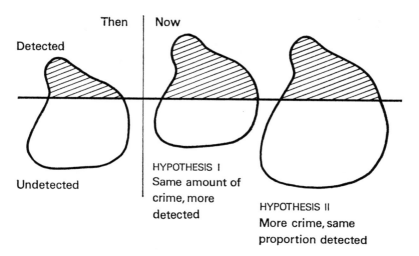

Figure 31 Two hypotheses to explain the observed increase in crime, based on the analogy of the iceberg.

ordinary, by gratifying sadistic impulses and by reducing the struggle between police and criminals to a sort of football game.'

Many speeches and much writing have been devoted to the hypothesis that crime on television produces crime in people, but it remains a hypothesis undeserving of the word 'true' used in the above quotation.

To emphasise the scepticism of scientific thinking it is instructive to enquire whether it is even true that the crime rate is increasing. A White Paper of 1964, *The War against Crime in England and Wales 1959–1964* states:

'The total number of indictable offences recorded by the police rose between 1958 and 1962 by 43%. The number of offences of breaking and entering rose by 47% and the number of offences against the person by 48%. Convictions of indictable offences increased between 1958 and 1962 by 39%.'

These figures can be presumed to be based on accurate observation of accurate police and court records and as such are facts, but note that they only refer to offences *recorded*. Crimes are rather like an iceberg, only a proportion is visible. It could be that the present 'crime wave' is because more of the iceberg is now seen, rather than because the iceberg has grown larger.

It is a useful exercise to find arguments that support the alternative hypothesis that a greater proportion of crimes are detected. (R33) Consider the three main stages which result in a criminal act being recorded in the annual statistics. First, the crime is detected and reported to the police, second the police investigate it and third, if a suspect is caught, the police charge him. Now consider the possibilities at each stage which could be responsible for an increase in crime detection.

1 The proliferation of private telephones has made communication with the police very much easier and quicker. People who might have been reluctant to leave the safety of their homes at night to report suspicious events from a public call box, would not be afraid to use their own telephone.
2 Constant references to increases in crime on television and in newspapers, and dramatised stories with scare headlines in the latter, make people more suspicious and consequently more likely to report unexpected 'happenings.'
3 Television programmes such as 'Dixon of Dock Green' and 'Z Cars' may have altered the public image of the police so that more people are inclined to co-operate in reporting suspicious events.
4 In recent years the police forces have become more mobile by the wider use of cars and motor cycles, and radio transreceivers have improved their communications with headquarters. Consequently the police can be on the scene of a reported crime more quickly than in the past.
5 The decision to charge an offender, or merely reprimand him with a warning, may be influenced by the extent of public interest in the crime. In particular this may be the case in juvenile crime; the policeman who formerly boxed the ears of boys throwing stones may be persuaded by public concern over juvenile delinquency to take the boys to Court.

Notwithstanding these arguments most informed people take the view that in general crime is increasing, but from a scientific point of view we should be prepared to treat this as hypothesis and not fact.

If there is some uncertainty about the extent of crime, there is considerably more about its causes, and even more confusion as to what to do about it. Perhaps this is only true of the expert, as is suggested by the following comment by a Home Office research worker (R34) on the causation of crime:

'It is perhaps surprising that with a subject about which the experts know so little, the layman knows so much. Ask any normal citizen about the causes of crime and it is certain that you will get a quantity of positive replies. Ask how offenders should be treated so as to ensure that they do not commit further offences and few will have any doubts about the methods they advocate, unless perhaps they have made a study of the subject. Degrees of certainty of belief in this field seems to be inversely proportional to knowledge of facts.'

In 1960, the Gallup Poll (R35) investigating punishment questioned 1 100 people who were claimed to be representative of the adult population of Great Britain. One question was: 'What should be the first concern of the Courts in sentencing a criminal?' Four possible answers were provided. The responses are shown as percentages of the replies.

To punish him for what he has done to others. 36%
To do what they can to reclaim him as a good citizen. 32%
To punish him to stop others following his example. 25%
Don't know. 7%

The Old Testament view of crime and punishment, 'an eye for an eye and a tooth for a tooth', is clearly prevalent. Of course, these responses are value judgments.

Another question revealed that 74 per cent of the sample believed that certain crimes merited flogging or birching; 79 per cent of those holding this opinion believed that the result of reintroducing such punishment would be a reduction in the appropriate crime rate. Whether a crime merits corporal punishment is of course a moral value judgment but the latter belief in its deterrent value is an unsubstantiated hypothesis. Years before this Gallup Poll the Cadogan Committee (1938) had stated:

'After examining all the available evidence, we have been unable to find any body of facts or figures showing that the introduction of a power of flogging has produced a decrease in the number of the offences for which flogging may be imposed, or that offences for which flogging may be ordered have tended to increase when little use was made of the power to order flogging, or to decrease when the power was exercised more frequently.'

and similarly in 1963 a report of the Church Assembly entitled *Punishment* said:

'Corporal punishment has not been shown to have any reformatory value and there are signs that it hardens attitude towards authority.'

The hypotheses that corporal punishment does, or does not, deter have been neither proved nor disproved—there are too few facts. The recent abolition of some forms of corporal punishment is a result of the changing moral value judgments of the community and legislature.

Punishment may have several purposes. It may be simply a revengeful act to repay a culpable act; or it may serve as a deterrent to the criminal himself or to potential criminals; or it may be an attempt to re-educate the criminal to the normal values of everyday life. The effectiveness of particular punishments as retributions is not open to scientific study, but in principle at least, the effectiveness of a deterrent and of re-education is measurable. Here again we find much opinion but little fact. Baroness Wootton (a prominent magistrate and social scientist) wrote (R36) in 1963:

'There is no way of knowing whether the sentences for which my colleagues and I have been responsible have discouraged or encouraged crime; and no way of knowing how our particular records in this respect compare with those of other benches. And what goes for us goes equally for the most eminent and experienced judge in the country.

If we are to take advantage of the help that science might give in the business of sentencing, there must first and foremost be a change of attitude.'

The causes and prevention of crimes are bedevilled by opinions and untested hypotheses. In the Home Office Research Unit, the Cambridge Institute of Criminology and the Universities about forty scientists are engaged on these problems at the time of writing. Not only is much more research needed, but those who will use its results—policemen, prison officers, magistrates, judges, probation officers—need to recognise that some of their opinions may be built on the sands of prejudice rather than the bedrock of fact.

Envoi

The scientific method of tackling problems has many facets, but above all else there are two outstanding precepts for problem-solvers:

1 Distinguish between the facts, the value judgments and the hypotheses.

2 Be ruthless in testing hypotheses against the facts.
These are the Pillars of Hercules of the modern world—the gateway to Truth.

Suggestions for Further Reading

S. D. Beck, *The Simplicity of Science*, first published in the U.S.A. 1959, Pelican Books edition 1962.
K. Stone, *Evidence in Science*, J. Wright, Bristol, 1966.
Two accounts of scientific method from a biological angle.

W. I. B. Beveridge, *The Art of Scientific Investigation*, first published 1950, paperback edition, Heinemann, 1961.
Descriptions of the use of imagination, intuition, observation and reason by well known scientists.

J. B. Conant, *On Understanding Science*, first published in the U.S.A. 1947, Mentor Books edition 1951.
An historical approach, based on case studies of scientific method.

R. W. Jepson, *Clear Thinking*, fourth edition 1948, Longmans,
R. H. Thouless, *Straight and Crooked Thinking*, second edition 1953, Pan Books.
These two books are concerned with true and false methods of thinking. The former has a number of useful problems.

J. Bronowski, *The Common Sense of Science*, Heinemann, 1951, Pelican Books, 1960.
J. Bronowski, *Science and Human Values*, Pelican Books, 1964.

References

1 *Sunday Times*, 16.12.62
2 This is a modern rendering quoted by N. L. Munn in *Psychology, the Fundamentals of Human Adjustment*, Harrap, 1961
3 *idem*
4 *idem*
5 *The Anatomy of Judgement*, Hutchinson, 1960
6 Figure 8 and 32 are taken from reference 5. The 'hidden man' was originally devised by P. B. Porter (*American Journal of Psychology, 67*, 550, (1954)
7 *Guardian*, 4.2.64
8 *New Statesman*, 1956
9 Quoted by Humby and James, *Science and Education*, Cambridge University Press, 1942
10 Quoted by S. L. Payne, *The Art of Asking Questions*
11 *Times Educational Supplement*, 21.2.64 and 28.2.64
12 *Guardian*, 18.4.63
13 *The Economist*, 27.12.58 (A humorous article on the misuse of statistics)
14 Suggested by J. L. Cloudsley-Thompson in *The School Science Review*
15 *The Observer*, 3.11.63
16 *Evening Standard*, 2.58
17 Quoted by Humby and Dowdeswell, *School Science Review, 34*, 15, (1952)
18 *New Scientist*, 8.4.65
19 Quoted by S. Glasstone, *A Textbook of Physical Chemistry*, Macmillan, 1948
20 *Smoking and Health, a Report of the Royal College of Physicians of London on Smoking in Relation to Cancer of the Lung and other Diseases*, Pitman, 1962
21 *Hansard*, 16.7.63
22 *Guardian*, 3.4.62, 11.4.62, 18.4.62
23 *The Conduct of the Fluoridation Studies in the United Kingdom and the Results achieved after Five Years*, H.M.S.O., 1962
24 Quoted by Drummond and Wilbraham in *The Englishman's Food*, Jonathan Cape, 1957
25 Quoted by J. Maddox, *Guardian*, 10.12.63
26 Leaflets published by the London Anti-Fluoridation Campaign, circa 1965
27 Mausner and Mausner, *Scientific American*, February 1955
28 *The Use of Dipped Headlights in Birmingham: a Report of the*

Lord Mayor of Birmingham to the Minister of Transport, 1963;
The Birmingham Dipped Headlights Campaign, 1962–3, Road Research Laboratory, H.M.S.O., 1963
29 John Downing, *New Society*, 11.2.65
30 Vera Southgate, *Educational Research*, 7, 83, (1965)
31 *Guardian*, 29.12.61
32 *The Times*, 3.1.63
33 J. Wakeford, *New Society*, 16.5.63
34 L. T. Wilkins, *Educational Research*, 4, 18, (1961)
35 *Capital Punishment and Corporal Punishment*, a Gallup Poll conducted for the *News Chronicle*, 1960
36 Barbara Wootton, *New Society*, 14.3.63

Figure 32   An interpretation of figure 8 on page 29
Reproduced by kind permission of The American Journal of Psychology.